T0137566

i

ALSO BY J. KANNEGIETER:

"DESTINY'S OBLIGATION"
A realistic drama on Earth and in space, set in the 29th
century, when typical human fallibility and selfish territorial
interests threaten civilization.
(Trafford Publishing)

OUT OF NOTHING ?

And the Perception of Certainty

by

J. KANNEGIETER

Order this book online at www.trafford.com
or email orders@trafford.com

Most Trafford titles are also available at major online book retailers.

Printed in the United States of America.

ISBN: 978-1-4269-3661-6 (sc)

*Our mission is to efficiently provide the world's finest, most comprehensive book publishing
service, enabling every author to experience success. To find out how to publish your book,
your way, and have it available worldwide, visit us online at www.trafford.com*

Trafford rev. 6/16/2010

 www.trafford.com

North America & international
toll-free: 1 888 232 4444 (USA & Canada)
phone: 250 383 6864 ◆ fax: 812 355 4082

To our grandchildren,
Kirstyn, Evan, Ava, and Markus

Two things fill the mind with ever new and increasing admiration and awe, the more often and steadily we reflect upon them: the starry heavens above me and the moral law within me.
(Conclusion of 'Critique of Practical Reason' - Immanuel Kant, 1788)

TABLE OF CONTENTS

INTRODUCTION
There is infinitely more than we can ever know!

'Out of nothing at all' is a repetitive line from a popular '80's song (by 'Air Supply') and, like most songs, it relates to love or the lack of it! But that is appropriate here, since love in every form is one outcome of the mysterious phenomena discussed in this book. It articulates and exemplifies our wonder about where it all comes from and what it really is, – what we sense and experience. Without resorting to philosophy or religion, we occasionally ask: why anything at all?

Writing this book is clearly a dumb idea since my first one was a spectacular commercial failure. However, I enjoyed doing it and another book seemed a good formula to amuse myself in retirement. "OUT OF NOTHING? - And the Perception of Certainty" deals with ideas and concepts collected over 50 years. It is a hodgepodge of subjects, of interest mainly because they are mysterious. It bothers me when the question 'Why' is responded to with 'That's how it is'. Although any analysis in conversation is considered boring, listening to opinions is often very interesting. We don't always remember the reasons for firmly held beliefs, and that is normal; we shouldn't scrutinize opinions too often. But we must be willing to change, because closed minds obstruct progress!

At first it seemed audacious to write about things that confound experts, and it postponed this book for years. In the end I needed something to do and hardly anyone read my first (science-fiction) book, so it didn't matter much. I also wanted to record some of these fascinating riddles before over-maturing, and no longer remembering or caring. The solution was to write a book that refined my sketchy observations and opinionated conclusions but was also of interest to others (although few illusions remain on that score). To be honest, it was written for myself, out of a subconscious need to know and to see where it would lead!

It is natural to be a bit jealous of young people in their early twenties; they are at a point in life where they must accept their genes, their training and life-experiences, and decide which road to follow. Of course, most never stop long enough to seriously consider their options. What was the impetus for this book? Only personal interest and the challenge! And to stop you from asking, I am not formally qualified for much of it. But who is, when the subject is the fuzzy boundary of human knowledge? It shouldn't be a deterrent anyway; the main objective was to satisfy my curiosity and, as far as this book is concerned, let the chips fall where they may! It is arrogant to have a secret motto, but it could have been: "Think big, and never forget how very small you are".

This is not a science or history book, and it doesn't quote many textbooks either. It accepts mostly what physical and social sciences tell us, illustrating here and there but concentrating on many obvious contradictions and important unresolved issues. It attempts to apply sound reasoning to a mix of theory, paradoxical facts, and unorthodox imagination, including my own. How current knowledge was arrived at is well documented and accessible to anyone via the Internet; there is no point in boring you, or myself. Reading this book again I will pick and choose from the table of content, depending on my mood; there is not much continuity between sections. It is recommended that you do the same, because reading something uninteresting is a waste of time. You can even start at the end, and subsequently discover what inspired such conclusions. We are all semi-conscious of the many intangibles in our lives, although we usually ignore them. Reading about such things provides an opportunity to contemplate (and criticize), although I have tried not to philosophize. Religion is a huge intangible but this book should not provoke anyone since it was written by an agnostic. My sympathies lie with any religion that does not insult our intelligence by ignoring the obvious.

Without resorting to dry academic dogma in science, psychology, religion, or philosophy, the themes of this book are -- Who and what are we, and where did we come from? -- How much insight do we have into how and why we make decisions? -- How do we relate to our physical world and what really is that? And -- Can we imagine the very long-term future beyond our concern for Nature and the many unfortunates among us?

It describes basic conflicts in Science using analogies, without mathematics! Controversial issues are presented to heighten awareness and provide perspective. The objective is not advancement of science but a better understanding of it. One exception may be a new approach to gravity in Chapter III, and that is allowed because no one knows yet what gravity is anyway. In general, the aim was not to improve or create expertise but to awaken a cognizance of what is still missing. Apologies are made for opinions that are naïve, pretentious, hypocritical or impudent, because conflicts between logic and instinct are always based on uncertainties, and any truth is only assumed. Contrary to what you may conclude, the socialistic slant is minimal because personal initiatives are clearly and unalterably an indispensable part of us. However, unchecked profit-taking by shrewd opportunists creates unbalances in society that jeopardize any government's initiative for essential minimum social equality.

Halfway through the book it became obvious that stability in Nature and our lives ensues from duality and cyclic balancing between opposing forces. For example, neither rigid capitalism nor communism can ever be a permanent social solution. Or, the Universe needs gravity to stop electro-magnetism from blowing it apart. It is incredible to me that 2000 years ago the Chinese already identified the principle they called 'Yin-Yang' as the basis for stability in every aspect of Nature, something not really emphasized in scientific literature to my knowledge.

Over the years it was interesting to see how much the opinions of experts changed, even on topics seemingly inviolate. That is normal, and it spurred the immense growth of social, scientific and technological knowledge. However, it is also a reminder that most knowledge is empirical (not fundamental, but based on experience or experiments). This is the broad book-theme; it asks awkward questions and in some cases discusses hypotheses that are hotly disputed by experts. I can do this because such experts will never take me seriously, since I am not one of them. Nonetheless, authority makes mistakes sometimes; there are usually many conjectures to any controversy, and sometimes they are all wrong!

Historically the science-elite always assumed the comfortable attitude that only a few wrinkles in their theories needed ironing out. Recent discoveries in various fields now confront theorists with serious contradictions hinting at 'hidden variables' in the foundations of science, and that phrase alone will cause some academics to throw this book in a corner! Hopefully it exposes such contradictions, making them more intelligible to myself and to anyone interested but lacking the expertise. Several hypotheses, supported only by minorities, are discussed and compared with orthodox views. The questions are of prime importance, but some tentative answers are interesting! Ideas are like magic; without logic applied to imagination we have only instinct and memories of what came before. Progress then depends entirely on biological evolution!

Professionals in every field of human knowledge are taught to reject unconventional theories, unless supported by overwhelming evidence. This is called the 'scientific method' and it protects Science from charlatans and chaos. Defying this unwritten rule jeopardizes any career in Science. However, history lists many crucial advances by individuals who, mostly with luck and extreme persistence, forced the acceptance of their deviating concepts. Nevertheless, students looking for a happy life are better off to play by the rules and defend established dogma. This book differs from most others by assuming that the reader is unfamiliar with accepted theory and disinterested in the details and arguments for its defense. Many paradoxes are presented in a simplified manner, making it more superficial but easier to understand and, hopefully, more entertaining.

Like me, you probably don't entirely understand what social and physical Science is all about these days; it is too complex, unless you happen to be an expert in some area yourself. Science has gone up the ladder of knowledge to great heights, but it still disappears into the clouds. This book has a different approach; it occasionally assumes end-scenarios based on informed guesses, describing it in ordinary language. This is contentious, and the reader adds as much as the writer because language only conveys ideas subjectively. I am an ardent fan of modern Science, not a critic, but it does seem that the attitude of "Certainty" assumed by ancient Greek philosophers, that everything is based on Earth, Water, Air, and Fire, has been substituted by its modern equivalents.

Right or wrong there are tentative answers to all questions, except for three big ones: - How did it all start? - What is behind it all? , and - Where are we going?-. Most sciences scorn untestable hypotheses; the scientific method embraces 'top-down' theoretical models that agree with experimental evidence, but only if it does not challenge established dogma. Such accepted models are sometimes incorrect or incomplete, with changes suggested occasionally by new evidence. Although difficult, important progress is often made by rogue visionaries who foresee an untestable hypothesis and develop it into a 'bottom-up' model that also agrees with reality and the experimental evidence. It illustrates the creative magic of the human mind and in all fields such visionaries have contributed to the elevation of knowledge. Magic is a

4

romantic notion, anathema to Science, but you will encounter some confounding issues that still have a touch of magic. Science anticipates their exposure soon, but I don't believe it. The intent of this book is to review interesting circumstantial evidence without boring you. My wife reads none of my writing; she says it all sounds like a business letter to her. Well, mush or circuity may be tried some other time, but not here!

This book is intended for non-experts; bona fide specialists are expected to put it aside after the first chapter. Actually, my hope is they ignore it altogether. Legitimate criticism is constructive but subjective appraisal is a downer. Experts will not be asked to review this book; they'd probably refuse and I don't want to embarrass them either. A few friends read an early draft (see back cover) and I am very much indebted to them; it gave me some direction! A common complaint was that the scientific information was hard to follow but, without real interest, concentrating is difficult! Science books intended for laypersons use lots of scientific names and references that block comprehension of all those convoluted theories. Such books suffer from too much information, disastrous for a poor memory, although most details are best forgotten anyway. Only concepts are important! I was much impressed by Paul C.W. Davies' book about the origins of life, called "THE 5th MIRACLE– The Search for the Origin and Meaning of Life" (Simon & Schuster), and delighted when he allowed me to quote a few excerpts.

'Out of Nothing?' addresses the complexities and uncertainties arising out of intelligent life, questions about science, physical or non-physical anomalies, and about human behavior and society. The sub-title could have been 'Making Sense Out of our World', but there are thousands of books attempting to do that. People often ask me what could be of interest to them, and I usually respond that it provides few answers but that it will increase their uncertainty! You'll encounter abbreviated life-stories of several very special people in history who made a difference, concentrating on why they made a difference.

Readers may recognize the contentious issues and view them from a new angle. It is a rather personal voyage of discovery and, like on any voyage, compatible company is very welcome. It is probably better not to discuss the esoteric issues, unless smirks on people's faces don't bother you. It is not 'cool', and will receive comments such as "That is too deep for me"! But that is the point; most people go through their very busy lives purposely avoiding the most obvious questions of existence.

CHAPTER I - <u>THE HUMAN CONDITION</u>

IMAGINATION & CREATIVITY
Imagination uses memory and intelligence to project the past, present, and future.

<u>Are Humans Unique?</u>
Speech, a large brain, and imagination make us special

It seems reasonable to start a book such as this by reflecting on imagination and creativity. Topics were shuffled around so much that in the end this section came first, almost by accident. If what you are reading is not interesting then take a look at the Table of Contents; it should always be your choice, not mine!

This being the opening section, let's start with a controversial assertion! Like everything else in this book it is premised on observations, deduction, and evidence adapted from many sources. What really makes us human, more so than our large brain, is the organization of that brain (i.e. our mind). It was the miraculous combination of language and imagination, supervised by intelligence and supported by memory and instinct that made us distinct. Appearing as a mutation about 180,000 years ago, this seems to have been one of the most significant and far-reaching step-changes in biological evolution since the central nervous system developed the first rudimentary memory. It was the magic spark, and the reason that humans can claim to be special without embarrassment. It represents an unbounded new force, since everything else in Nature is restrained by the counteracting forces of stabilizing cycles. Human intelligence, linked with supporting artificial intelligence, is limited only by an inability to perceive what is at the very bottom of it all!

Some evidence for the above assertion may be anthropological artifacts from Neanderthals, cousins to a human predecessor species. They also had a very large brain, memory, and intelligence, but they lacked in the imagination and speech department. They lived successfully in Europe for 200,000 years without ever changing their customs (food, tools, or habits). But when our species (Homo Sapiens) arrived there, 30,000 years ago, they were quickly pushed aside and became extinct. Intelligence, compassion, aggression, and creativity defines humanity but it is the unique gift of imagination and social communication that provides consciousness, allowing us to have a sense of who, what, and where we are, in the present, past, and future. We should not downplay the importance of speech because that was an essential prerequisite for imagination, or the ability to categorize, project, and remember what will or has occurred in the future and in the past. Language means assigning labels (words), and a mind without language is as useful as a 20-volume encyclopedia without an index.

It is a strange question, but does a dog (or dolphin, or any other intelligent animal) have imagination, or language? They have to some extent, but nothing like us; for them emotion, instinct, and memory determine everything they do, based on what in the past provided food, comfort, or security. A dog uses short-term logic, such as 'how do I get there?' They are curious opportunists and, although anything new is accidental, the experience is retained in memory. Cats are experts at sneaking up on prey by using the environment, and that is instinct and learned behaviour. Put a sheltered dog in heavy city traffic and the emotional overload will probably cause panic and death. Do the same with humans and they'll take one look at the cars flying by and never venture off the sidewalk. That is suspicion and fear of the unknown (instinct) and intellect (logic) combining! We'd be lost without our imagination, condemned to a life like a cat or a dog looking for food when hungry, sleep when sleepy, and no real planning. Mind you, some people would be happy with that! Despite creationists, it is obvious that our intelligence, speech, and imagination are unique characteristics that developed gradually through natural selection and mutations from a specific sequence of mammal species over more than 100 million years.

Few uncertainties in our life are more influential than instinct, the urges and predispositions we wonder about sometimes. Many people don't like the word, it reminds them of ancestral animal connections. It is very recognizable in the newborn, because babies are helpless for a long time except for the noise they make when hungry or uncomfortable. Some animals are born ready to run, swim, or hide, depending on how quickly they must adapt to stay alive. The newborn of some hoofed animals keep up with the herd amazingly fast because their brains are preprogrammed with all the necessary information, and they know to be quiet, or the wolves will find them. In contrast, human babies make a lot of noise and require a very long learning period before they can even crawl around, suggesting that instinct may be less of a factor in their survival. It may also be that human brains require this long to develop after birth because a baby's skull cannot be any larger; mothers would have difficulty walking.

A mystifying example of instinct is the Monarch butterfly. The North American variety migrates from Canada to Mexico in the winter but it is only the third or fourth generation that actually completes the round-trip. The question is, how do they find the exact place where their ancestors wintered the previous year? At least salmon were born in their mating destinations, as were most of a flock of migrating birds. They probably have a built-in clock and sense the Sun, or the direction of the Earth magnetic field; but for Monarch butterflies there has to be more. Their instinct must provide them with local markers that are a permanent feature in their tiny little brains. Of course, that misrepresents the case because their DNA string is just as long as ours. This is the key to instinct; they (and we) are born with an enormous amount of pre-programmed instructions, seated in every section of DNA that directs physical brain development, and evolution put it there! Thank God (and evolution) for intellect and logical reasoning that allows us to choose between intuition, memory, and imagination, otherwise we'd be pre-programmed robots, like butterflies!

When people are unable to imagine themselves in someone else's shoes (like a sociopath), or cannot distinguish fantasy from reality (psychosis), these are mental deviations. The first is caused by a lack of imagination and compassion while the latter suffers (among other things) from excessive imagination, unchecked by reality. Evolution provided humans with functional combinations of instinct, memory, and intelligence (logical thinking), allowing a projection or assessment of the future based on experiences from the past (despite common debilities like depression, phobias or addictions). Imagination is essential for the management of any situation by recognizing a similarity to past experiences or learning. It is a spontaneous process, using memory and intellect to direct preprogrammed motivations. Imagination is curiously unreal, the stuff intuition and dreams are made of, and we are usually better off not to think too much about where it all comes from. We usually try to stay on solid ground, having suffered the consequences of emotional decisions when logic was obscure or unpalatable. Unfortunately, logic may also go out the window when in love or after drinking too much!

Pondering about what makes us say or do things reveals how complicated our motives are, not at all based on the cold facts we assume. This does not condemn our decision-making processes, far from it! Anyone making only logical decisions, without any emotion, may be an ultimate social criminal when also managing to act obligingly. Such people appear to be outstanding citizens but it is only an act to hide their selfish motivations! Keep your eyes on those with leadership ambitions and faked compassion, they are intelligent self-promoters with a hidden agenda and there are a lot more of them than you know!

Creativity and Benefit to Society
Memory, logic, and imagination assist the creative instinct

Creativity draws on imagination and memory by identifying or suggesting opportunities, and that is why evolution advanced it. It enables us to assess the suitability of any new arrangement. Creativity is an ingredient in everything we do, such as speech, manual and mental work, science, technology, art, play, and social interaction. It is satisfying to be creative, it imparts subtle motivation, but we also admire creativity in others. The 'Renaissance' period in 15th century Italy is a good example of how creative encouragement and unimpeded imagination awoke the entire western world. It is easily stifled at an early age because society rewards conformance. Even children's play is now overly guided because computer games and toys generate profits, or because an ambitious soccer coach insists on dictating game-play. You'll be more successful in today's world by learning, copying, and conforming than by emphasizing your creative side. Still, many people pay the price and some have the talent, intelligence, and emotional stability to succeed by filling a genuine need. They are the drivers and the innovators who rise above this pressure to conform. Ironically, in our society the really successful people are those who exploit such innovators; they put up the money and take the risks, and that is also creativity.

The urge to be creative has a genetic basis, but it is frequently suppressed by formal training stressing conformance to established processes and standards. Motivated to be original, some people revolt and tenaciously pursue new ideas, but often at their peril. If there is a common trait in very creative people it is their aversion to authority and they ignore instructions from everyone. On the other hand, innovative people willing to conform can be very successful if they are emotionally stable and blessed with intelligence and a good memory. They recognize and judiciously utilize the creativity of others and their ego does not compel them to advance their own ideas. Such social success demands specialized competence, an agreeable personality, motivation, and good luck (being at the right place at the right time).

People in the 'genius' category are often obsessed by a need to match some image they have of themselves, and nothing else matters. Their creativity is then a means to achieve subconscious objectives and successful only if it fills a need in society (or amuses people). Others, enlightened or misguided, choose not to conform because they dislike existing yardsticks, and they are the rebels who receive a lot of undeserved media attention. The vast majority of people achieve their creative satisfaction by improving on what they were taught, and that expertise keeps society moving going!

Something new always triggers curiosity, a learning and survival-trait all animals have; but human imagination and creativity goes beyond that. Formal education at a young age is essential because it trains the mind and transfers historical knowledge that provides a background of facts instead of myths. Any social group teaching myths and ignoring factual evidence is bound to end up in the intellectual doldrums. Evolution equipped humans with a very powerful simulator and we store in our conscious and subconscious memory the images of locations, shapes, and circumstances of experiences since early childhood. We enjoy cross-connecting such memories to events we have not actually experienced (fiction-novels are a good example).

'Imagination' was crucial in our ancestors' struggle for survival and it is therefore not surprising that evolution has advanced it. All people have the ability to imagine creatively, combining remembered experiences, intuition, symmetries (order and beauty), and logic; and we use it constantly to secure and retain our place in society. Some people are amazingly creative in certain fields and they stand out. The history books are full of people who managed to impress others with their superior creativity. Millions of others may have been equally creative but were unsuccessful because there was no significant benefit to society. They became frustrated for a while, until some other creation held the promise of a pot of gold.

Inspiration and Major Advances
Advances in civilization were social adaptations
of creativity

Creativity is what rejuvenates society. However, creativity is often useful to an individual without actually benefiting society. Carving little elephants for sale to tourists is creative and beneficial to the carver and the local economy but does not advance society much. It requires deliberate motivation and inspiration to select socially useful objectives. The exquisite religious statues in some old cathedrals are artistically similar to little elephants, but they meant a great deal more to society over the years, inspiring millions. Creative thought with the objective of future benefits is what elevates humans far above other life on this planet. It demands motivation, imagination and intelligence, advanced by multi-million years of evolution and paid for by the fatal misfortunes of billions of our ancestors.

Everyone has creative thoughts, sometimes brilliant, but few have the opportunity, motivation, and perseverance to see it through. When it changes society there is a hint of magic, suggestive of some purpose behind it all! But it demands dedication and singlemindedness bordering on obsession to pursue an idea that stands out among millions of others. Revolutionary ideas do not just pop up one day; it takes many days, separated by weeks, months, or years, until the original idea is analyzed, modified, and no longer subject to rejection. If you want an example, how about Thomas Edison and his light bulb? And what would motivate anyone to choose such a tortuous and uncertain path? Psychologists should answer that, but it probably is a natural instinct to be resourceful and make life less precarious for yourself and inadvertently for society!

Creativity and inspiration are very interesting to me, working for many years in a semi-research technical environment and surrounded by a mix of talented and well-educated experts. On one side of the technical staff were the creators, incapable of convincing themselves that their ideas were off the wall sometimes. On the other side were those who could recall so many facts that according to them anything new could never be successful; they remembered too many failures. All our management needed to do was form mixed teams and let the dreamers convince the cynics; and that worked, at least some of the time! The innovators never overloaded their memory with facts, leaving room for optimism and creativity. The inevitable controversies were interesting and enjoyable.

Inspiration is triggering your imagination and creative thoughts into new and practical patterns. It usually occurs by chance and completely unexpected. Concentrating too much on a single problem fixes thought-patterns and closes the mind to intuitive inspiration. Virtually all major advances started with sudden insight, when several concepts and experiences combine in a novel manner. Intuitive judgement that one approach is better than another is not a hunch but a semi-conscious assessment, balancing necessity against possibility. Artificial Intelligence (A.I.) will never do this, unless it is programmed to roughly assess many provisional models and then compare suitable candidates based on the latest relevant information. Creative thinking requires looking at a problem from the top down and bottom up, in detail and in overview, all at the same time. Such a filtering process needs instant information, an ability computers excel in but they lack flexible logic (to select something not the same but only similar). Will A.I. ever acquire this human skill to accept a hunch as a temporary fact, and ignoring the lack of information simply by assuming what is most probable? My guess is that they will and that the human / A.I. partnership will carry us throughout the Milky Way in a few milleniums.

Art and Science of Da Vinci
Excellence is a statistical occurrence in any large
population

A supreme example of creativity, ingenuity, and talent emerged in the late 15th century with the Italian Leonardo da Vinci (1452-1519), born near Florence. His reputation as a universal artist, scientist, and inventor was (and still is) such that he has been called 'the Renaissance Man', and 'Homo Universalis', or 'the most talented person to have ever lived'. His innovations changed painting-techniques forever and his science notebooks and technical inventions inspired millions of people over the following centuries. It is therefore ironic that his impact on society was overshadowed by something as vulgar as the introduction of firearms and cannons on the battlefield, which occurred during his lifetime. We cannot say that Leonardo was opposed to it because he happily participated in the design of new weapons and the defenses against them. Suddenly castles, swords, shields, and even horses to some extent were no longer as important. It completely changed the conduct of wars and thereby the course of history.

Leonardo was a charismatic man, presumably homosexual, who managed to lead his life without making any known enemies. He was never short of wealthy admirers who commissioned him for paintings, architectural designs, sculptures, design of warfare machinery, hydraulics and other engineering projects. It was reported that he died in the arms of his friend, the French king (Francois I). He and his younger contemporary Michelangelo are unrivalled in the history of art,

13

although Leonardo was invariably a procrastinator. He could always find reasons why something was not yet finished. This is why his (13000 page) notebooks were important; he did not feel the need to go back to every entry. None of his sculptures or architectural designs were ever fully completed; he probably did not have the patience nor was he ever satisfied, but they do affirm his genius.

It is therefore amazing that Da Vinci managed to stay on friendly terms with his sponsors, although he exasperated every overseer accountable for his progress. His knowledge overwhelmed everyone, based on immense powers of observation, intellect, and personality. His notebooks show clearly what was of interest, and it includes some very advanced (for that time) concepts of anatomy, engineering, optics, flying machines, and in particular new techniques of painting. He never had any formal training in science or mathematics and some scientists of that period were jealous of the excessive attention he received. His greatest contribution to civilization are the few (~15) paintings that survived, because they represent the summit of all art. His 'Mona Lisa' and 'Last Supper' paintings are the most famous of any age.

<u>Natural Talent for Art</u>
Art's origin is our instinctual desire for order,
symmetry, and beauty

Art is a prominent intuitive expression of imagination and creativity, although art appreciation differs immensely between people. Art cannot be defined in absolute terms; it is the product of imagination and creative force, and entirely personal. Some of it is acquired and one of my pet peeves is that schools teach art without really making kids like it. They end up copying others instead of drawing on their own imagination. We had an interesting perspective sketching class in Secondary School; you were not allowed to use anything except a pencil without an eraser. Being good at it, the teacher asked if I would consider an illustrator career. Of course, I was only interested in sketching airplanes and other such things, but it was still art! Some children have a real natural talent and nobody can stop them. However, what is beauty for one is not necessarily beautiful to anyone else. We can look through books of famous old paintings (or modern, for that matter) and have little interest in any of it, although the craftsmanship may be superb. And yet, occasionally we'll see a print of something and it is immediately obvious what the artist wanted to convey. It creates a mood, a message, an excitement, and you keep looking at it. Many art snobs are eager to spout opinions that make them sound superior. People feel intimidated because famous art often does not appeal to them and they wonder what is wrong with them, but there isn't. Any creation can be called art, and it is usually interesting, but you don't always have to like it. Art is personal, and it is good art if you like it. The same goes for music, and probably even more so.

Music is a uniquely human experience; it correlates with patterns in your memory and synchronizes with it. You anticipate what follows and it is a part of you. Music can be hypnotic and therapeutic, especially when everything else is excluded. It relieves stress and makes you feel good. Even animals like a melodic repetitive tune; it probably relaxes and puts them to sleep. When listening to a violin concerto or a pop tune it plays in your mind, but only if you recognize and like it. Some people get goose pimples from listening to music that is special; the effect can be that powerful. Many children never acquire a real appreciation for music, except what happens to be popular with their peers. This is probably because all music requires some conditioning and exposure. They cannot bear to listen very long when it lacks familiarity. Musical geniuses are very different; and their brains are unique. Mozart composed entire pieces of music in a flash, able to write down an entire score in outline without ever stopping. This is inspiration in a pure form. He is reputed to have produced an illegal copy of a closely guarded piece of music, after attending its performance in Rome's Sistine Chapel. The process of creating can often be enthralling and exhilarating. We can imagine how Dostoevsky felt after completing a novel, or J.S. Bach while listening for the first time to the opening movement of his Orchestral Suite #1 overture. Or da Vinci, after painting the Mona Lisa (no wonder he never wanted to finish it). Or James Watt when his steam engine actually worked. We don't usually think about it, but that feeling is the summit for creators! And monetary reward is never the only motivation.

Mozart, a life of Genius
Musical genius is sporadic, narrowly focused,
and exceptional

We should make a detour, back to music and Wolfgang Amadeus Mozart (1756-1791). No composer ever received more attention, although maybe not during his lifetime. His years as an amazing child prodigy are well known and we won't dwell on it, or the details of his life somewhat fictitiously portrayed in the movie 'Amadeus'.

Contemporary composers such as his friend Joseph Haydn and a young Ludwig van Beethoven were entirely in awe of him. He influenced the latter's music greatly, but Beethoven's plan to study with Mozart unfortunately never materialized. Although admired by many, and a local hero in Salzburg, the general public was less impressed with Mozart's music. Even the Austrian emperor offered the opinion that it had too many notes! Since Mozart was a contract musician he had to compose to the liking of his employer, the Archbishop of Salzburg. His work during that time was technically supreme but nothing like what it could have been, if motivated differently. Music is mathematically based, with scales, pitches and octaves, all referring to sound frequencies and their ratios. Many mathematicians are also good musicians but Mozart was different, a supremely creative composer-musician, who could develop a new composition in his mind and remember it. He eventually quit his job, lacking fulfillment, to freelance as a keyboard musician and composer in Vienna and to marry Constanze, a singer. He was quite successful, especially with operas he was not allowed to do previously. His friendship with Haydn was very influential in later compositions.

Expressing my own opinion, after this preamble, Mozart's early music is excellent as church music, or for 18th century dance, easy listening, relaxing, or what have you. It is technically amazing and mentally therapeutic, but it does not hit me like some of his later compositions, sections of his operas, and particularly his Mass in C minor and the Requiem (unfinished at his death). The composer suddenly switched from being a master technical artist into an incredible genius, which he had been all along but never allowed or willing to fully unleash. In my opinion (!), some of his later music surpasses Beethoven in emotion and Bach in subtlety and creativity.

If insight into creative genius and a life of inescapable chaotic emotional conflict is of interest, you should examine the (sketchy) information regarding the circumstances surrounding Mozart's death (and please forget the movie!). Probably suffering from acute rheumatic fever, he frantically outlines and partially details his 'Requiem'. Someone commissioned it after losing his wife, intending to publish it as his own work. Mozart accepted only because of the very large sum of money offered. However, suspecting he is being poisoned and fearful of his own approaching death, he becomes obsessed with it and works day and night until Constanze takes it away. There was a strong rumor that on the day before his death she gave it back to him and he requested a performance in his bedroom by musician and singer friends. Reaching the 'Lacrimosa" part he became severely distraught and the performance was stopped. After his death there was a big row over the publishing rights. Real life drama easily surpassed the artificial turmoil of his operas. Mozart was a freemason and therefore his funeral was boycotted by the Catholic Church (the reason for his well-publicized pauper's funeral).

SOCIAL HISTORY

The world may know, that so far as we approve of monarchy, that (in America) THE LAW IS KING. For as in absolute governments the King is law, so in free countries the law OUGHT to be King; and there ought to be no other.
(Thomas Paine, 'Common Sense', 1776)

Colonialism and Independence
Western colonialism inadvertently encouraged global homogeneity

There are few examples that better describe social evolution than the demise of colonialism in the 20th century. Pursued by Western nations as a means to improve their status and economic competitive positions in previous centuries, colonialism was tainted by slavery, racism, native poverty, ruthless exploitation, military disparity, and intimidation. Often, the natives were seen as sub-human and inferior, soothing the exploiters' conscience in the unlikely event that was a concern. Colonialism fed on all the negative aspects of human motivation. It is therefore surprising that it actually strengthened the social base in many of the exploited countries by making the entire world more compatible. One reason was of course that their original autocratic rulers were often just as bad, if not worse.

Western empires were eventually forced to give up their colonies due to a variety of circumstances; but none did it voluntarily. Communications had improved, people were better informed, and Communism was a threat everywhere. Old empires had weakened or disappeared and the end of the Second World War was a new beginning, with America as a somewhat self-serving champion for

17

liberty. You don't need higher education to know that foreign domination is unethical. The colonial powers quickly discovered that sending soldiers to oppress the oppressed had become very unpopular after the Second World War. Although independence has not been the expected panacea for many of these new countries, and much support is still required, the road is now clear and social evolution will decide.

Gandhi and Nonviolent resistance
Social negotiations and restraint are essential to reduce violence

No one is identified more with the demise of world colonialism than Mohandas Karamchand (Mahatma) Gandhi (1869-1948), the leader of India's national movement for achieving independence from Great Britain by means of non-violence. During the mid 20th century he became a symbol of freedom for oppressed people all over the world. Born into the merchant class (caste), he married at 13 and finished his education as a lawyer in Great Britain in 1891. Unable to open a successful practice in Bombay, he accepted a job with an Indian company in South Africa. Here he was not only confronted but also subjected to racial discrimination for the first time. His proud nature set him on a course fighting for basic civil rights of Indian immigrants in South Africa, continued in India when he returned in 1915.

He earned the respect of India's peasants and laborers by organizing protests against domination by higher castes and governments and its crushing hardships. They gave him the honorary title of Mahatma ('Great Soul') and it stuck with him for the rest of his life. He advocated the principle of 'Satyagraha' (force of truth), a name he coined for his doctrine of nonviolent resistance and non-cooperation. It was based on ideologies used in various religious uprisings, and became popular in India. When Great Britain decided in 1919 to crush public unrest with brutal force, resulting in slaughter, Gandhi publicly called for a general strike by nonviolent non-cooperation and passive resistance. This was eventually successful and the British were forced to back down and release Gandhi from jail. This set the tone for his relationship with the colonial and other authorities for the next 25 years. He, his wife, and several followers spent many years in jail over that period.

Social, religious, and economic pressures in India at the time were confusing, especially for foreigners. The British took the easy way out by propping up existing aristocratic power structures. They were afraid of Bolshevist influences and large-scale revolution. India's large population (>600 million) was split into so many different interest groups that control was perplexing. Following his campaign of non-cooperation violence erupted despite his pleas for restraint and Gandhi was arrested and sentenced to 6 years in jail. Released after 2 years for health reasons, he realized that the cooperation between Hindus and Muslims had badly deteriorated in the meantime. The poor and non-influential were a large majority and they supported Gandhi and the Congress Party, now led by his collaborator Jawaharlal Nehru as President. The British responded in 1935 by giving Indians autonomy over provincial governments. This did not stop the unrest and violence erupted often between Muslims and Hindus, ignoring Gandhi's calls for nonviolence. He adopted a very simple lifestyle, attempting to connect with the lowest in Indian society by wearing their typical clothes and eating their diet. He realized that nonviolence was far more effective in his tightrope act of social, economic, and religious inequality. It made it difficult for the authorities to silence or arrest him in full view of the international media. His attitude towards other religions was that all had basically the same message and that love, compassion and nonviolence, together with the old Greek Golden Rule ("do not to others what you do not wish for yourself"), was a common universal basis. He criticized the cruel Indian caste system and particularly the plight of millions of 'untouchables', but his target was always Britain.

During the Second World War, Gandhi was sympathetic with Britain's struggle but insisted that India could not provide support except as a free country. Their 'Quit India' campaign resulted in violence and mass arrests on a very large scale, again resulting in his arrest, his wife, and the entire leadership of the Congress Party. His wife died in jail and Gandhi was released towards the end of the war. The Labor Party in power in Great Britain after the war favored independence for India and provided assurances that allowed Gandhi to delay the struggle, triggering the release of many political prisoners. The Congress Party voted to accept British terms to partition the country into Hindu-India and Muslim-Pakistan over Gandhi's strong objections, but it avoided the threat of a religious war.

Complications arising out of splitting the country caused utter confusion, prompting Gandhi to start his last 'fast to death'. It was an attempt to force agreement between Hindu, Muslim, and Sikh leaders, and avoid all-out war. He got his way, but on the 30t[h] of January 1948 Gandhi was assassinated in New Delhi by a Hindu radical who blamed him for compelling the Indian government to make partitioning payments to Pakistan. He is remembered everywhere for his nonviolent political and social reform, and in India as the 'Father of the Nation'. He was voted runner-up to Albert Einstein as 'Person of the Century' in 1999 by Time Magazine. However, he never received the Nobel Peace Prize he clearly deserved because it was considered too controversial.

Vested Interests and Laws
Vested interests are central to Business and Politics

Our acclaimed democratic institutions have always been potential conveyances for ambition, vested interests, protectionism and other negative motivations. Nothing is perfect you'll say, but political elections could be based on merit if excessive spending and secret promises to contributors were outlawed with enforced punishment. Of course, the media is then in charge, and that presents its own problems. Communism didn't really take off in the twentieth century because it undermined personal motivations and exposed the instinctive selfish corruption of power. Removal of international restraints (immigration, trade-barriers, etc.) may cause a chaotic free for-all, hampering the execution of power, and it will never happen voluntarily!

Why has the United Nations so little clout? Ah, but vested interests favor that! Pragmatic sources tell us that what we have is better than nothing; but all well-off nations maneuver (at least in essence) to weaken the UN's authority. This house build by idealism is now occupied by pragmatists. Nations with everything to gain from a strong UN are playing destructive politics, reducing their influence, and helping the powerful to maintain control. Survival instincts clearly overpower common sense, and that is nothing new! Political evolution will never overcome it; so our civilization may have to destroy itself (several times?) in the next 50,000 years, until biological evolution revamps our genes. Nature has a lot more time than we do!

Despite all that, nothing contributed more to order and wellbeing in human society than the concept of democratically established laws, administered by an executive subject to the same laws. It separates the last two hundred years from the rest of history, and gives meaning to the phrase 'equal rights'. Of course, it depends on the ability of our elected legislators to vote correctly, and not be influenced by selfish interests, religious bias, or other discriminations. On the other hand, minorities should not stubbornly (or violently) insist on inordinate consideration due to unfair treatment of their ancestors in history. Ancestral rights and freedom of religion are a matter of the mind, not society's debts that carry over to following generations indefinitely. Never mind what happened in the past, in a socially oriented democracy all babies are born equal. Most people can go back in history and find incidents during wars, etc, where their ancestors were cheated and therefore claim compensation for loss and suffering. It's a big, bad world, and the only pragmatic approach is to support democracy and fair laws on the largest possible scale.

Benjamin Franklin and Democracy
Franklin's pragmatism and humanism
enlightened the world

It is a pleasure to highlight Benjamin Franklin (1705-1790) as one of my heroes and an individual who contributed greatly to critical changes in history by his role in the American Revolution. That uprising served as an early example of democracy, civil rights, enlightenment, and free enterprise to the rest of the non-democratic world. He must be viewed relative to his contemporaries, many of whom were preoccupied with the gentlemanly custom of protecting vested interests such as slavery, favorable tax conditions and privileges of class for the rich, economic isolation, and male social and political elitism. Although old and in a feeble state, he was one of the most important contributors to both the U.S. Declaration of Independence and Constitution (as President of the State of Pennsylvania) and was probably the most well known American in the western world at the time.

Franklin somehow negotiated timely and crucial financial and military help from France, making Washington's victory over the British possible, and also directed the peace-treaty with the British after the war. He should be ranked second only to George Washington in his importance to the founding of the United States of America. His influence, popularity, and communication skills provided a spark to the subsequent spread of democracy and the demise of aristocracy in Europe. He influenced many of his prominent friends in France, particularly the initial leaders of the revolutionary 'National Party' (Bailly and Lafayette) in the French National Assembly. The following is an excerpt from his speech on the day that he signed the US Constitution and it is an inspired expression of his views and priorities:

"I confess that there are several parts of this Constitution which I do not at present approve, but I am not sure I shall never approve them. Most men indeed as well as most sects in Religion, think themselves in possession of all truth, and that wherever others differ from them it is so far error. I doubt whether any other Convention we can obtain, may be able to make a better Constitution. For when you assemble a number of men to have the advantage of their joint wisdom, you inevitably assemble with those men, all their prejudices, their passions, their errors of opinion, their local interests, and their selfish views. It therefore astonishes me, Sir, to find this system approaching so near to perfection as it does; and I think it will astonish our enemies, who are waiting with confidence to hear that our councils are confounded like those of the Builders of Babel. I hope therefore that for our own sakes as a part of the people, and for the sake of posterity, we shall act heartily and unanimously in recommending this Constitution wherever our influence may extend, and turn our future thoughts & endeavors to the means of having it well administered".

Benjamin Franklin was a very accomplished individual, a widely known American author and printer, politician, scientist, inventor, statesman, diplomat, and anti-slavery proponent towards the end of his life. Before and during the signing of the U.S. constitution he served as President of Pennsylvania for three years. He was famous worldwide for his inventions in electricity and other things such as bifocal reading glasses. Typical was his proposal to prove that lightning was electricity by flying kites in a thunderstorm. This experiment was repeated by many and some died when the wet string conducted electricity too well. His pointed lightning rod became a standard throughout the world. He also discovered and experimented with refrigeration by evaporation after noting that evaporation made you cooler in a wet shirt than a dry one, especially on a hot windy day, and resulted in his publication 'Cooling by Evaporation'. Earlier in life he printed and published 'Poor Richard's Almanack', writing much of it himself (under pseudonym) and it became very popular. He also founded the American Philosophical Society for the review and discussion of science. His inventions and wit made him very popular in upper class Parisian society for the seven years that he lived there, especially with the ladies who insisted on inviting him to every social event. This was useful to him politically and helped to secure vital French concessions for America.

We can make a comparison with Michael Faraday who was born the year after Franklin died, in 1790. Both were schooled only in reading and writing, but they educated themselves by working in the book and printing trade. Both were curious about the mysteries of electricity, with Faraday sort of picking up where Franklin finished off. They were unusually ethical and logical thinkers, but on a different planet as far as religion was concerned. Faraday was a firm biblical Christian although opposed to any established Church; Franklin was a deist, accepting God but unbelieving in any scriptures or prophets. One of his many public sayings was: *"The way to see by Faith is to shut the eye of Reason!"* All in all, Benjamin Franklin was a worthy Hero who did not need much advice from anyone and who managed to avoid upper-class comforts and triviality. In the end, his most important and lasting contribution to humanity may be the influence he had on acceptance of basic human rights in western civilization.

Territorial Instincts and Survival
Only reproduction is more potent than our territorial instincts

Human civilization is a good example of how order is created out of apparent chaos, but only an extreme optimist will say that all is well! Prehistoric migrations exposed small and isolated groups of people to the rigorous stresses of survival. Many useful human character traits and social innovations are traceable to the isolated occupation of new territories, although ruthless exploitation or annihilation of native populations or wildlife often accompanied it. Biological evolution increased order by selecting human-traits that optimized survival of individuals in small groups. Intelligence has been our ticket to survival for the last million years. Elephants had size, lions had teeth, and birds had wings, but we outsmarted them all.

It still applies; if you follow your instincts you'll drink too much, drive too fast, and you'll die! In recent history, kings or dictators and their inner circles increased the survival-rate of ever-larger groups of humans by emphasizing territorial, patriotic, religious, or racial animosities. The history books will tell you that it was usually at the detriment of their neighbors. Evolution advanced our motivation to seek social alliances for security, but the execution is adaptation and learned behaviour. It is the reason for our huge and complex society favoring alliances for all individuals, ranging from families to towns, schools, clubs, and workplaces; drop out and you'll be part of a different species!

People feel protected within a family, a village or a nation. Group aggression is mostly directed at foreign, racial, cultural, or religious factions and recent wars have killed millions. Is this also progression of order? Well yes, political organizations have become more regulated than they used to be, but the cost in human life was more than ever before in history. Although not a factor in evolution, it does bring to mind Hitler's ideas of creating an orderly society! In time someone's Ph.D. thesis should tell us how much the obsession of a few caused large-scale human suffering simply by stirring up prejudices in a majority.

It is ironic that we as individuals can be happy and content while at that same moment people perish elsewhere under the worst possible circumstances. Modern communications overwhelm and numb us with such news. Our compassion switches off in protection and it makes us seem more callous than we really are. Human life has always been this way, out of necessity, and entire civilizations have disappeared to make room for new experiments. Individual survival is paramount, and we are social creatures only for as long as it suits us. Compassionate logic suggests that the U.N. should be reorganized to intervene always, to smother wars and genocide when it surfaces, or reduce the impact of major natural disasters. Instead, it is mostly tied down by insufficient funds, lack of a clear mandate, and by politics. Repeatedly we see reporters talking about such events as their assignment of the week, mixed-in with other news about sports, gang murders, etc. The efforts by a large number of charitable agencies are commendable but they usually lack the clout and resources to be very effective. Although contributions by individual nations are haphazard, it still shows the potential of what a fully coordinated U.N. effort could accomplish; but who would get the political accolades?

Leaders and Individuals
Successful leaders may cater only to majorities,
but it is immoral

Unchecked imagination often steers us in directions we regret later. Although everyone makes mistakes, it is a moral offense against family, society, and ourselves to consciously deviate from common sense, compassion, and society's laws. Surprisingly, it is often those with so-called conservative views initiating such suspect actions. Many people (top leaders among them) argue that in some cases future society will benefit from strong unilateral actions harmful to a minority of innocent

24

individuals. This may be so, and people never tire of quoting historical examples. But the truth is that vested interests, and not society's future, are always the real motivation. Actions that hurt innocent people are morally improper. A serious attempt to protect them must come first and only then can it possibly be justified, assuming that it does prevent greater harm to a majority. Failure to do so should be classified as illegal, and a war crime in serious cases! This is heresy to many, and it will be dismissed as the ignorant opinion of a weak-kneed pacifist. I'll go along with pacifist, but not weakness. Pre-emptive actions are what we should be suspicious of. Although inevitable, it always seemed that a more humane target than Hiroshima and Nagasaki could have been selected at the end of the war with Japan, thereby minimizing the death of mostly innocent victims. Unfortunately, and typical, these victims were not emotionally perceived as innocent at that time!

Notwithstanding all this, what should an individual do when expected to carry out tasks that are morally dubious? Should we volunteer for active duty in an ideological war that does not directly threaten our family or community? Should we participate in a firing squad, when ordered to do so? We usually cannot change choices already made for us; and it would be entirely different if we could. Assuming that we cannot influence the outcome of top-level decisions, the choice is personal and my sentiment would be to participate because not doing so is unfair to others. But we don't need to publicly applaud it in the name of patriotism! The bottom-line is that genuine threats to society cannot be ignored, and the Second World War serves as a useful example. It is a warning to any group with lunatic ideas of grandeur that the rest of the world will not for long close their eyes and hold their nose. Eventually enough steam collects and leaders will be motivated to make the necessary decisions. Ethics and civility are then ignored, until the malignancy is removed from society.

The proliferation of nuclear weapons has changed and severely complicated the methods of dealing with hostile groups or nations. Nuclear blackmail could be particularly effective for terrorist organizations and, if that ever happens, the only reasonable response is to freeze out all their support without mercy. It is the time to circle the wagons and swallow national pride by pursuing every possible alliance, insisting on execution of meaningful rulings by international courts. The arrogant disdain and non-support for such courts by major powers today may be sorely regretted in times to come! Their citizens typically assume immunity against international aggression, deeming its invulnerability strong enough to protect them. Individuals in dominant countries often feel superior and that creates a false sense of security. Conversely, people in countries with low status feel inferior and they actively resent it. The bond as members of the human race ought to be stronger and more permanent than shared but accidental citizenship! Such instinctive artificial division must be diminished, and not strengthened as is the norm, in order to solve our world's biggest problem!

SOCIAL EVOLUTION

The force behind society is social instinct, shaped by self-interests, compassion, and common sense.

Instinctive behavior and Logic
Logic and instinct are our masters, and we pay
attention to both

Instinctive behavior is clearly more influential than we are aware of or will admit to. Advanced by evolution and supervised by logical intelligence, our survival and procreation instincts implicate every subconscious thought, action, and social interaction. The intricacy of this precludes a detailed explanation, but a physical basis for it must reside in our genes. Anyone who does not believe that should carry out a little experiment by asking someone what he or she would like to do at that moment. Go fishing? Do they eat the fish? No, but they like to catch fish; it is a challenge and exciting. Fishing and hunting is part of our survival programming and anyone enjoying it is more likely to survive. Think of any activity you like to do and it'll have some instinctual footing, including paintball war-fare games!

Free will based only on logic is a fallacy, unless we become aware of all the emotions motivating us and deliberately override them (the modus operandi of religions). A society doing this openly and logically for altruistic purposes may be a Utopia to wish for, although it inhibits our natural motivations! We have the potential for it, because many individuals in recent and past history have lived their lives without ever compromising altruistic principles. Their life-stories are of great interest. Without mentioning the obvious ones, such as the originators of various religions, many philanthropists have secretly helped others, without any obvious benefit to themselves. Of course, we don't know, they may have been afraid of burning in hell! Alternately, you can override compassion single-mindedly to achieve selfish objectives and that is 'free will' also; it can be socially positive or negative! Talking about hell reminds me of a church notice; it said:

At the evening service tonight the sermon topic will be 'What Is Hell?'. Come early and listen to our choir practice.

It is easy to select examples where instinct dominates logic. Any extraterrestial intelligence unfamiliar with Earth biology can easily discover how humans acquired their instincts by simply following evolutionary logic. Even the reproductive practices of biological organisms on Earth have a logical basis; it happens to be the most efficient way to mix and spread genetic variety. Yet, the complexity concocted by nature to motivate organisms to mate may still be very puzzling to them. They will view it as strange and probably amusing that female peacocks admire colorful and intimidating feathers in a mate. An attractive face and body-shape lures human males although fully aware that a compatible personality will be far more important for lasting love and marriage. The females of higher animals typically select mates who can dominate others, with a body exhibiting good health and strength. Romantic love is the emotional pinnacle for humans, an obsession, and the subject of songs, movies, books, etc. It is clear that logic plays only a minor part here. It is all very entertaining, and there probably is a lot more puzzling behaviour non-terrestrial visitors would shake their (multi?) heads over!

We could not function without the evolutionary wisdom of instinct and its offshoot, intuition. We learn as children to question it, control it, or override it, but we cannot not do without it. We learn with age that a happy life depends on judging it to suit the circumstances!

Cultural heritage and Superstitions
Beneficial habits and insecurities have formed
our instincts

The latest spontaneous mutational boost in human intelligence (Homo Sapiens) seems to have occurred about 180,000 years ago in Africa, possibly in a single individual. Rudimental speech, useful for group survival, evolved earlier and that may have been the trigger that made higher intelligence effective. Even an intelligent leader cannot benefit the tribe without communication! The rules of natural selection and genetics passed on this exceptional mutation without dilution to a gradually expanding fraction of descendants. It is fascinating to imagine what that must have been like. Did they try to fit in, or did they outwit their less intelligent neighbors? The differences were large enough to discourage much intermixing, and it created separate tribes substantially different from others.

It would be interesting to know what this new species thought about their amazing and bewildering surroundings, because they were now substantially more capable of contemplation. However, natural selection had yet to evolve this larger brain into what we now recognize as uniquely human characteristics with a supporting body structure, adapted for language, reasoning, and cultural traits and an extensive social lifestyle. Without this, and without the accumulated heritage of knowledge collected and passed on over thousands of years of observation and deduction by billions of our predecessors, it must have been overwhelming to say the least! They had to believe that something put them there, and it filled their lives with superstition and wonder. It is not difficult to imagine yourself in such mystifying circumstances; little kids experience it, convinced there are monsters under the bed!

The collective human psyche eventually became thoroughly conditioned by illusions providing some rationale for the inexplicable manifestations of nature. Today we know (more or less) the real causes for things such as lightning, rainbows, mirages, tornadoes, tsunamies, earthquakes, fire, the sun, and the stars. Put yourself in their place! You must admit that supernatural explanations are the most obvious. They were primitive in our eyes mainly because they lacked knowledge and not because they were physically or mentally inferior, although they may have been to some extent. A capacity for supernatural belief evolved as an instinctive part of them because you functioned in such a group only if you participated in its customs and lore (analogies). If not, you became an outcast who would not survive and your DNA was not passed on. Such evolutionary pressure is where our superstitious nature and instincts come from! Today, Science clarifies how nearly everything functions, although the question of how it got started and where it will end is not close to being answered, or worried about very much!

Free will, Self-esteem and Lifestyle
Motivation and social interaction is usually
analyzed belatedly

Human autonomy (free will) has been a serious question for all philosophers ever since Socrates. Classical science allowed no loopholes for events to occur outside of the deterministic laws of nature and its initial conditions. However, there is a good reason for assuming you are our own boss, because you are! Humans do not have a robotic brain, it operates instead on Quantum Mechanics' 'Uncertainty Principle'; we take huge shortcuts and make firm decisions based on dubious assessments, feelings, and fuzzy mental pictures. We decide which restaurant to go to by recalling our satisfaction from the last visit, the expected cost, how far it is, how busy it will be, and most importantly: (in my case) where does your wife want to go (and she cares about the ambience and how well they do chicken!).

We should feel sorry for people in the future who must check with robotic consultants for everything; it will never work and take all the pleasure and surprise out of life. Such A.I. advisors may also require a 'fuzzy brain' mode! We encounter this today when computer analyses provide many options to choose from, but it still up to us to consider the intangibles. Jumping ahead a little to my speculative ramblings about physics (just in case you decide to miss that!), we are incredibly fortunate that Nature has a dual personality in everything. For instance, electro-magnetism and gravity independently affect us. It means that neither one can lock us into a rigid system. It is difficult to understand the philosophers' problem because 'free will' clearly develops from the conflict imposed on us by two masters: logic and instinct. And just ask: we always act logically, if only everyone did!

Behavior of people is interesting. You don't need to be a psychologist to recognize that self-esteem is a prime motivator. Without knowing it, people act to project an image they have (or want to have) of themselves. Conflicts occur when this image is challenged and pride is hurt. Some fortunate people are very secure and never argue over anything inconsequential; they have learned to step away. You have to like yourself to feel secure, blemishes and all. Of course, some people like themselves far more than they should and still get the benefit (which is hardly fair!), and a few end up despised leaders of society. Most people consider themselves secure, but they still act with either indifference or reverence towards others, depending on any domination or intimidation they sense; but that is self-esteem! History's saints showed us that peace of mind is accessible to anyone, by never consciously doing anything detrimental to others and by deliberately avoiding pretense. Self-glorification is a poor substitute for respect in society but almost everyone engages in it at one time or another. It can become an overriding urge, to appear better than you really are, but it is not a good path towards a happy life and it usually backfires.

It is little things in life that make a difference; by themselves insignificant, and we tend not to notice. Recently my wife and I walked along a quiet back street in a Mexican town when a window-washer came out of a small office building with his bucket and tools. He looked up and gave us a big smile, automatically. We smiled back and said 'Allo', imitating the Mexicans working in our hotel. The difference here was that this man did not need to acknowledge us. He had no motivation other than to express his satisfaction with life and lack of preoccupation with weighty matters, unlike everyone else on that street. He could have ignored us; instead, he made us feel good. Today, in urban places at least, casual relationships with strangers have radically changed. As a boy I remember complete strangers talking to me, asking me what I was doing, or where I was going, or just about anything else as a sign of interest. It made me feel noticed and recognized as a person. Few people do this now, probably for good reasons because it may be misinterpreted.

People today are not really less considerate than they used to be. As a shy teenager in continental Europe the degrading rudeness when lining up, or fighting for seats on a bus or train is clear in my memory. People now live more hectic lives, with instant and constant communication, swamped with information, duties, gadgets, passwords, problems, and responsibilities; and it collectively inflates into stressful uncertainty. We are programmed to regard idleness as bad; we need money to pay for everything, including mini-holidays due to hundreds of stressful commitments, and smiles fall by the wayside! My stepfather told me with some disgust that workers in Java, before the 2nd world war, worked just long enough to pay for one month of rice, etc, and then quit to relax or work for themselves. I could never figure out what was wrong with that! Of course, I had visions of sleeping under palm trees, blue oceans with white sandy beaches, etc.

The technical revolution had a big impact on lifestyle and behaviour, although it is difficult to recognize now without some contemplation. For example: we all view photographs or video movies from our past occasionally and these pictures are almost as familiar as the people we know today. It serves as a direct memory link and helps to refresh information that might otherwise become clouded. It is easy to recall what our kids were like nearly forty years ago and that makes a difference in our relationship, for me anyway. People in much earlier times did not have that and their recollection may not have been as good. We can now retrieve information for almost anything, and it has had a huge impact on business and personal life. The arrival of cell phones and other gadgets is a little dubious; it removes the protective bubble of privacy we all need to stay on top of the water, not below it! And it did not made things any easier. All aspects of life today are affected by competition, from pre-school until old age. It stimulates people and the economy, but the stress eventually wears some people down. Changes in lifestyle should be cultivated, not forced. Expectations by young people to be a manager one year after they leave school is symptomatic of today's impatience and it suggests a lack of respect for the experience of older people.

It is risky to complain too much about young people, for fear of being called a grumpy old man. The challenges they face are different from what I remember, although not necessarily any greater! Prosperity and easy credit made it easy to spoil kids, and that is something to guard against. Then again, we shouldn't generalize because 50 % of the world's children live in severe poverty, with 25000 dying from it each day, and that is by far the most shocking statistic we ought to be concerned about! My gripe against young people in the Western world is that they keep changing our classic Christmas songs, not just the words but the tunes also. It bothers me! Sing them as they are, or compose your own!

Social and Political Philosophies
Social instincts advanced collective security of communal life

There are many people today presenting a casual image to those around them with a cool 'I don't care' attitude. Many young people adopt this behaviour, and it may be immaturity or a lack of self-esteem. However, it could also be a subconscious recognition that the need for social status is just another instinct. It is obvious that in pre-history the lowest on the social totem pole was also the least secure. Today, our improved social and economic security has reduced the need for high social status. It was no accident that the 'hippy' generation grew up after the depression and second World War, never really knowing what inescapable hunger or hardship is. However, quality of life still depends almost entirely on success in gaining the respect, trust and affection of others. Therefore, we are motivated to compete for status among ourselves. Social disparity in earlier centuries evolved naturally out of varying abilities between individuals. Smart, attractive, stable, or charismatic people often married into families with more possessions, becoming richer themselves. This selection process became less important in Europe after the French revolution, although its initiators had no such grandiose motives; but the timing was right!

Could aristocrats not see how unfair life was; that it only leads to disaster when you enslave others to claim a disproportionate advantage? They must have, since they considered themselves much smarter than the rest of the population. However, there is no such thing as a smartness gene, it depends on too many random combinations; they just steered the system for the best possible ride. Mind you, they really believed themselves superior, which was understandable because the unwashed rabble did not have the benefit of education and were good at deceiving and stealing (no wonder!). The higher classes ranked the lowest classes of society only a little above their dogs and horses. The whole situation was a textbook case for social natural selection and survival of the fittest. However, smart is not good enough, you have to remain strong! The fittest eventually turned out to be the mob, eliminating aristocracy in most places except where group sentiment and a resistance to change allowed them to exist in a symbolic role. The key to any revolution is always timing and coordination.

Pragmatism defeated idealism in the early twentieth century, after workers and Bolsheviks overturned Russia's aristocratic rulers, mainly in protest against the enormous casualties during the first World War. Raw violence then forced that country into a dictatorship, culminating 70 years later with another revolution that rejected hard line communism. This was also predictable, because common sense equality is not the same as some intellectual ideal; it must be practical and compatible with social instincts. Everyone in the Soviet Union knew that equality could never happen, that it was only a slogan. Executed more democratically it might have evolved into the sort of democratic socialism we now see in some other European countries where it serves as a useful comparison with democratic capitalism.

Critics of modern capitalism charge that insufficient basic social provisions makes wage-slavery inevitable. Proponents of capitalism believe that it is the only natural formula for maintaining social health, and assure survival of the fittest. The question facing us with ever-increasing complexity in this new millennium is if improved social equality is more important than individual opportunity. The obscene excess of inequality inherent in a purely capitalistic system is clearly unhealthy. On the other hand, social equality like communism stifles individual initiative. Sustained political stability and progress is possible only when both objectives are looked after. This is not surprising since everything else in Nature is the result of a balance between two extremes.

Any kind of global government needs to satisfy our human need for social equality, freedom, and individual opportunity in order to be successful in the future. Our evil instinct for territoriality and an 'us against them' attitude must be curtailed at all cost. After a century of upheaval and recovery we are once again at peril. Civilization proceeds in an upswing of complexity, causing millions of individuals to get lost and some nations to be overtaken and marginalized, or worse. This is neither extreme nor unusual and it will swing back like all evolutionary phases. Except for crumbling architectural ruins and environmental corruption, society's heritage of laws and cultures may be the only long-term enduring features in humanity's existence. Even that dwindles when too many individuals feel they no longer belong, triggering a new age of individualism. There are few signs people care; they'd rather spout left or right wing dogma or sarcastic satire!

Considering how things are organized, it is to be expected that business and financial leaders will pull every legal trick in the book to increase profits. It is the responsibility of elected officials to balance the interests of society against vested interests by ensuring that adequate regulation and directives are in place. There is no question that government owned or managed businesses are a bad idea; the need for profit is missing because the taxpayer will cover it. Competition in business, with reasonable rules, is what keeps it all efficient and balanced. However, secret deals, collusion, and plain deviousness of big business drags it effectively down to monopolies that rip-off the public, warily tolerated by governments afraid of losing party sponsors or upsetting the apple cart. The amount of shady money spent on elections and by lobbyists to influence the government is often enormous and obscene! Any solutions or recommendations? No, not really (or, may be, yes, start again!).

The Pitfalls of Education
Education means teaching of facts and the exposure of myths

Educating, training and motivating children has to be one of society's most rewarding tasks, although it takes a lot of dedication and patience. Detecting innate talent in each child, be it social, physical, intellectual, artistic or creative is satisfying, because each one has something special that can be cultivated. Many educators and parents judge a child by their relative performance, leaving a majority feeling inadequate in some way. Each child should be made to feel special once in a while, to develop its self-esteem. Many children do not perform well when asked to memorize things they don't really understand. I still remember the sinking feeling in early elementary school, trying to memorize a seemingly infinite number of fractions. One day my stepfather, trying to help me, became annoyed and said: "Why can't you remember that ¾ is 0.75, the same as 3 divided by 4"! That was the end of problems with arithmetic; boys simply don't pay attention. My recollection of childhood is still vivid, and it is a big part of me today. Despite many depressingly stressful and embarrassing situations or periods, climbing one hill after another is remembered most; anticipating what lies beyond with the exuberant optimism of youth. In old age you know what is on the other side, but you still look for new hills, just in case.

Many children develop roadblocks in learning because they do not expect to understand and are embarrassed to show it. Approaching new concepts from all angles requires imagination, so that even the dreamers in a class will get it. There are many stumbling blocks and teachers are not always equipped to recognize them. A good example is the simple subject of 'work' in a High-school physics course, and it baffled me for years! How many students (or teachers) realize that holding up a heavy weight with a stretched out arm is work, although it is not classified as such by physics rules (without displacement). Of course it is work, our muscles use up a lot of chemical energy doing it. However, a bridge holding up all that weight does not perform work; and atoms in molecules do not get tired because electrons can only change their orbit (and therefore molecular structure) in specific energy-steps, and if they let go it is called metal-fatigue. This simple explanation is sufficient to prove to all doubters that the outrageous and mystical concepts claimed by Quantum Mechanics are very real!

Interested students use their imagination but they just don't get it. Forced to accept and remember ambiguous facts that don't make sense is frustrating. Uninterested students are better off, they just remember what the teacher tells them. The above example of ambiguity in knowledge is easily avoided by telling students to study more physics before they'll understand it. Teaching mathematics is initially simple; add two more cows to three and you have five, and all children can follow that, using their fingers. Multiplying or dividing is more difficult because that involves learned or remembered procedures, not just a concept. Fundamental to logic and mathematics are the abstract concepts of proportionality, probability, and rates of change, among others. Repetitively exercising learned procedures instead of concepts usually gets first billing when things get complicated and that is when understanding goes out the window. You don't learn to drive by reading a car-manual

The Future of Society
Concern for descendants is illogical, but
evolution advanced it

In my opinion(!), modern society is not well served by excessive patriotism and territorial prerogatives encouraged everywhere. The Olympic Games are an obvious example although by no means the only one. It can be divisive, and often causes me to cheer for the underdogs. This has nothing to do with the athletes themselves, only the motivations of overzealous supporters. Intended to stimulate friendship in the world, human emotions (and ignorance) often push it in the other direction, reminding me of celebrations at the end of a bloody war. The only way to stop wars and animosity is global equality and tolerance, as unrealistic as that sounds. Friendly competition is healthy anytime and a good way to establish relationships; but membership in the human species has priority over national, race, or religious allegiances. Civilization and our long-term future will depend on it.

Human society, today and in the past, is by its very nature competitive and adversarial. Growing up in such an environment teaches you to fit in and participate in what is sometimes called the 'rat race'. That is learned behavior because basic instincts only push us to survive and procreate. People in big cities grow up appearing to be unfriendly, while those in small outlying communities seem more sociable. Therefore, our approach to relationships is learned behavior and personality. There are many ways to be successful in society, and it depends to a large extent on our strengths and weaknesses. Some people excel at learning from experience. Some have a photographic memory for facts, without necessarily knowing their origin, relevance, or accuracy. Others relate everything to imaginary situations and attempt the application of logic for answers to their questions. The most useful human aptitude is intuition, being able to separate fact from fiction and sense from nonsense without too much study.

Political orientations of governments are left or right wing, godly or godless, regional or nationalistic, or anything in-between. Prosperity is possible with all, but citizen participation will vary considerably. Which is best? It depends on whom you ask. Entrepreneurs condemn pure socialism and laborers dislike capitalism; a bit of everything is typical, but how much varies a lot. It should not be a theoretical issue; what matters is that it functions smoothly, that opportunities exist for everyone, and that a minimum standard of living is provided even to unfortunates who are unable to earn it themselves. Mentally handicapped or addicted people living on the street is unworthy of any society, regardless of a country's political policies. Anyone walking through such a neighborhood and believing they have no responsibility for it is simply callous. Everyone deserves some happiness in life! The best political approach will be decided by social evolution, assuming that this chaotic world can get its act together without blowing up.

What will human society be like, thousands of years from now? It is possible that part of the answer will be found far from Earth! Colonization of space is mostly fantasy and of little interest right now, but that will change. The information technology explosion combined with new energy-sources and a revolution in physics will make life in space not just practical but inevitable, just wait and see (not literally of course!). Colonies in space, divorced from Earth, will augment human evolutionary diversity like Darwin's finches in the Galapagos, although not necessarily for the better! If this sounds like Star Trek, it is not supposed to; I am not a fan!

HUMAN EVOLUTION
Intellect and a social instinct combined to account for Human Society

Achievements of Homo Sapiens
Speech and imagination evolved to enhance
human security

The elevation of Homo Sapiens ('knowing human') from older (extinct) species apparently occurred more than 150,000 years ago as a chance mutation in a small isolated group in Africa. This mutation was a larger brain with a very much-increased ability to imagine past and future events, situations, and circumstances, by using logic, memory, and instinct. Natural selection induced many adaptations to accommodate this new brain since then. We tend to assume that people in earlier centuries such as the Middle Ages (~AD1200) were not as smart as we are today. This is largely incorrect; they were equally capable and could have functioned normally today, like anyone raised in some undeveloped region of the world. The disparity was caused by the more primitive social system, unsophisticated lifestyle, and the minimal knowledge transferred between generations. In other words, today's society, science, and education make the difference!

People living thousands of years ago were capable of magnificent achievements such as cathedrals, ships, and literature, requiring a relative level of experience, skill, financial commitment, imagination, and willingness to take risks that equals projects undertaken today. It shows how much society has changed that such major feats were then accomplished with only the resources of either the Church or the aristocracy. Collective initiatives by the general population had to wait until the rise of independent commerce or modern governments in much later centuries. Before you begin to worry, this is not a history-book and mentioning the past serves only to show its relationship with the present or future. Nonetheless, history is magnificent and interesting, especially when you ignore all the moronic things that happened, and concentrate on why it happened!

Evolution and Genetics Theories
Evolution is the preferential selection of
beneficial mutations

Modern Evolution-Theory does not depend on natural selection only. Major step-changes are caused by random mutations (DNA errors) triggered by cosmic radiation or other environmental stimuli, although natural selection favors beneficial changes to eventually become entrenched in the population. Evolutionary biology holds that Natural Selection molded humans for maximum procreation in a constantly changing environment. Such criteria did not prepare us to be masters of the world, and much of what we are is in excess or wrong for that role. Philosophers and biologists today are pre-occupied with the question: What is Life and how did it arise? Physics and chemistry can also provide some insight because life is not like a machine; it takes countless unique chemical processes to form an organism. A large number of very complex molecules must work together, and such processes need time and precise direction. Continuous (i.e. classical) energy theory in physics cannot account for chemistry or biology and only the weird theory of Quantum Mechanics supports a life-process concept in principle. And how it got started is anyone's guess!

It is interesting that Charles Darwin (1809-1882), the famous English naturalist, carefully avoided the term 'evolution' and mainly talked about 'natural selection' and 'survival of the fittest'. However, he was well aware that random variations and mutations within a species were essential for natural selection to function. Also, he downplayed the subject 'Origin of Man'. There were many reasons, all relating to his social class and the religious attitudes in England at the time. He was intellectually confident about his evolution theory but very insecure about its impact on society and religion (and himself). Regardless, Darwin was successful in convincing a critical number of scientists to at least consider his ideas, detailed in his classic book 'The Origin of Species' published in 1859. His friends in the Royal Society enthusiastically did most of the persuading and promoting because Darwin suffered from poor health in later years.

Modern 'Evolution Theory' today is the scientifically definitive theory of how natural species evolved (although unspecific about origins). It is equally based on Natural Selection and the Genetics Theory, discovered by the Austrian monk Gregor Mendel in 1865 by studying the inherited traits of peas but not publicly recognized until 35 years later. The concept of 'genetics' was therefore unknown to

Darwin. However, Mendel probably did read 'The Origin of Species'. Darwin and his contemporaries believed that 'blending' passed on hereditary factors, but this left in question why a specific beneficial change should not eventually diminish; overpowered by existing populations. Mendel provided the answer with his discovery that traits and characteristics are inherited whole, or not at all. His experimentation and insight initiated modern molecular genetics, which says that individual genes (relevant sections of a Deoxyribonucleic acid or DNA molecule) are responsible for all specific inherited genetic traits. Offspring receive a random 50% of some 25000 genes from both parents, who only pass on one of two different genes for each trait they themselves inherited from their parents.

It is surprising that healthy humans have a natural life span well beyond their average reproductive years and even beyond their apparent usefulness to society. You would expect survival criteria to have curtailed their average life span to about 50, instead of nearly 80 years. Clearly very old people are still useful to human society, unlike non-social animals who typically die soon after the age they cannot or do not want to care for offspring anymore. Humans are not alone; elephant matriarchs control a herd and often live for a long time because their knowledge and experience is valuable. It is clearly beneficial to society when young people in reproductive years have elders around for stability, wisdom of experience, and as role models or assistants in rearing children. It is common for elders to control native tribes because younger leaders are more likely to make unwise and destructive decisions. This supports the notion that social conditions greatly influence life expectancy.

Nature and Society
We will always depend on Nature for the
necessities of life

Mendel and Darwin's contributions to science did not change society like some other scientific innovations, but it had a great symbolic and religious impact on how humans viewed themselves in relation to nature and society. Although still frowned upon, it was no longer immoral or illegal to be an atheist; and scientific acceptance of evolution theory provided some legitimacy to hundreds of semi-philosophical subjective arguments. Caricatures such as Darwin with the body of a monkey actually helped emphasize the physical similarity between humans and primates and challenged the prevailing concept of humans uniquely made in the image of God.

Someone else would eventually have proposed a theory of natural selection and, in fact, Darwin was forced to publish because Alfred Russel Wallace (1823-1913) had independently developed a similar theory at the same time, albeit not as far reaching. Darwin received much recognition during his lifetime, including a state funeral when he died. Mendel died an obscure abbot of a monastery because science at that time simply did not comprehend the significance of his discovery. Darwin was a product of his super-confused ('Victorian') period, and it interfered with his scientific conclusions to some extent. He did address (in 1872) the evolution of human culture, race, and psychology, concluding that Man still shares many traits with animals at lower levels. His conclusions gradually converted him from a devout Anglican to an agnostic, throwing up his hands in bewilderment.

Most of Darwin's concepts were developed and based on fieldwork he carried out in South America during his famous five year 'Beagle' voyage, returning in October 1836. He openly expressed the opinion that all human races were biologically identical, but refrained from commenting publicly on the religious or social ramifications of his theory. Others were not similarly restrained and the subject 'Evolution' became, and still is, a topic of great controversy (although it shouldn't be!). However, it served to defeat the arrogant 19th century European upper class myth that a disciplined lifestyle, behaviour, and education had created inheritable superior characteristics in their offspring. Genetics eventually proved that genes are too complex to be altered this way, although clearly a child of two tall people has a good chance of being tall also.

Nature's invention of DNA
Predicting heritable traits from ancestral genes is convoluted

Genetics has been in the forefront of medical research since the identification of the DNA molecule by Watson and Crick in 1953, resulting in the eventual sequencing of a human genome in 2003. It now includes the new and still relatively undeveloped field of 'epigenetics', which means 'above the genome'. It involves (for instance) tiny chemical tags of carbon and hydrogen that attach to the enormously long DNA molecule, without interfering with the sequence itself (the 'genome'). Cells and organs of a single organism differ from each other because genes (DNA sections) have to be selectively turned on (or 'expressed') via intermediaries to make specific proteins and this is the job of such tags on the DNA sequence, or the 'epigenome'.

These instructions are inheritable but, unlike DNA itself, they can apparently be changed by external influences such as environment, food choices, and lifestyle. Therefore, when one of identical twins develops cancer it is by no means certain that the other will develop it also. In fact, research has shown that the epigenomes of identical twins diverge with age, especially when they have a different lifestyle. Genomes are inherited and unchangeable, except by mutations, but changes in the epigenome seem to be influenced by an individual's life experiences, among other things. Even more significant, such changes may be passed on to our children or grandchildren! Although still very controversial, it suggests that a parent's personal responsibility extends well beyond childcare and training.

Medical disorders afflicting humans vary widely, but few have an entirely genetic basis. The first distinction in the classification of illness is that it may be either infectious or non-infectious. Infectious diseases are caused by pathogens (micro-organisms) such as viruses, bacteria, fungi, and parasites. Non-infectious diseases are for instance cancer, heart, genetic and also mental disorders, and they may have a wholly or partial genetic basis. Some infectious diseases caused by micro-organisms are contagious and passed on through contact by hand to mouth, water, food, sex, water vapor, insects, etc. Death resulting from any disease it is called 'death by natural causes'.

Biological growth and the Genome
Each individual has identical genomes in every cell of its body

The complicated and poorly understood process of biological growth may be crudely compared to the process of assembling a new gas barbecue after you take it home. The factory has design-information (the genome) for all of the parts and provides you with assembly-instructions (the epigenome) and all the part-numbers (genes). The instructions tell you how many of each part number are required and where (i.e. how many of which gene must be 'turned on' in what location of the organism). It is clear that medical science has only scratched the surface of identifying what triggers many health malfunctions and what possibilities of control may exist. In any case, the little blurb on epigenetics illustrates the staggering amount of information contained in every sub-microscopic part of our body. We should sympathize with Fred Hoyle (see chapter II), and his comment:

-"The notion that not only the biopolymer but the operating program of a living cell could be arrived at by chance in a primordial organic soup here on the Earth is evidently nonsense of a high order."

He was not a biologist and was therefore ridiculed, but that did not remove the conundrum. Although Science is convinced that it is all based on chance, there is no clear evidence either way.

Like Newton, looking at his thumb, it impresses me that my two feet are exactly the same size, although that is controlled by information contained in every cell of my body! Science tells us the origin of that is not a mystery! Supposedly they mean that you cannot run properly with different size feet because the lions will get you and you won't be passing on your defective genes! Atheists may have glib answers to it, that nature's laws and evolution make it inevitable. However, it is difficult to close your mind to all uncertainties. Like them, I also assume that evolution is responsible for the fact that Monarch butterflies know to return to the same hill in Mexico every winter, after spending their summers in Canada. But how do they manage that, when only fourth-generation descendants complete the roundtrip? And what about someone's cat, left behind when moving hundreds of miles, still finding its way to an unfamiliar home?

EMOTION AND MOTIVATION
Judgment of people should be also be based on their importance to others!

Emotion and Instinct
Emotions and life-experiences clarify instinct
and motivation

Emotion, or the anticipation of it, is our everyday prime motivator. We crave and dread it; it hurts and helps, and we embrace or purposely evade it. In short, emotional motivations makes us what we are and how we act, including all those 'cool' people and the rest of us labeled as affectionate, heartless, hotheads, romantics, or any other tag. The perils and opportunities of life affect our mental state and our body's reaction to it. Asking people what gives them an emotional 'high' (or low) results in so many different answers that it clearly touches our essence. Most often mentioned is love, of course. But there is a whole gamut of other reasons, ranging from births, deaths, children, arguments, playing or listening to music, careers, winning the lottery, relationships, promotions, creative joy, surprises, sport victories, unexpected encounters, and even violent computer games! It includes anything and everything that motivates, stimulates, excites, diverts, angers, or unnerves us.

Never mentioned voluntarily are the negative aspects of emotion and motivation, such as hatred, revenge, ego, jealousy, selfishness, and greed. One of the worst sometimes is pride, although it is usually regarded as positive. Pride in moral actions is a powerful positive motivator and a source of happiness. However, hurt pride in relations with others is destructive, and individual pride is the main source of arrogance. The basis for emotion is entirely instinctive, with intelligence and logic as backseat drivers only. If we never examine and challenge our motivations we may end up wondering why life did not turn out the way we hoped.

Why not leave personal subjects like emotion for a psychologist or a relations-councilor to worry about? Because they are interested primarily in people with problems and here the interest is how it affects society! We assume that achieving objectives makes us happy, but such satisfaction is only half the story! We also crave those rare moments of complete happiness with no apparent rationale. In other words: entirely emotional. It can happen anytime, even in times of stress. Something makes you feel good and all tension disappears, as if nothing will ever go wrong again. Of course, it doesn't last and things do go wrong, but you don't forget! And you subconsciously search for what made such a difference, without ever finding it. Some look for it in alcohol or drugs, although everyone knows that can be a one-way street to despair and depression. Others are convinced it has its origins in victory (or love), and they become obsessed with that. A more fitting answer may be inner peace or tranquility, with compassion and humor as close associates. Some lucky people are always happy with a humorous approach to life, although tragedy or difficulties happen to everyone.

Is it really necessary to become despondent before you seek tranquility? Experts claim there are benefits in questioning suspicious motivations, even when you're winning the battle; it may not be worth it! Take a walk for no good reason; talk to people deemed not worth speaking to before, or let others lead and don't push anyone into accepting your point of view. Most of all: be important to companions in life; happiness is not luck. All this is nothing new of course, and much easier said or written in a book than done! Many people find tranquility through exercise, meditation, yoga, or breathing and other procedures that reduce mental and associated physical stresses. This is the best approach; temporary stress is vital to achieve objectives, but will ruin your life when it accumulates excessively! Stress build-up is habit-forming and unnoticed until it debilitates mental functions.

Breathing exercises use a natural body-rhythm to induce pressure and release cycles that relax stress in a part of the brain associated with specific muscle-tension in the body. It works, although what works for one may not work for someone else. It requires experimentation, and most people give up on it too early, judging it unworthy of their time. However, done formally or informally, it achieves the same stress-reduction as jogging 5 miles, in much less time and with no wear and tear on the knees. And it is excellent advise to take a 'deep breath' (or several) and close your eyes before attempting something you are fearful of. You'll be too busy thinking about doing that! The trick is to coax the brain into preventing or reducing unwanted tension. Stress-overload often originates with mind-abuse from excessive abstract thought instead of simply opening the mind to the immediate and the senses. Abstract thought entails a lot of uncertainty (the source of worry), and sometimes it is better to just watch the world go by.

We are blessed with superb intelligence. Unfortunately, at crucial moments it is far more satisfying to let your emotions take over, unwise as that may be. There seem to be no guidelines, other than 'think before you jump'! On the other hand, it would be a dull world; and what would Hollywood do if no one ever flew of the handle? Just the same, you won't have a peaceful life unless you identify what makes you mad, before you get mad! Feelings of unfair treatment, deceit, rejection, etc, can lead to overpowering and obsessive hatred and, in extreme cases, violence. It is not very original to point out that such conflicts are caused by people unwilling to admit (or fail to grasp) that their feelings and actions are causing them a lot of harm. Vested interests are always involved, coupled with one-sided and self-serving convictions.

The most dangerous fallacy by far is 'what I do is sanctioned by God, so it cannot be wrong'! This arrogant and self-righteous nonsense killed millions in history. It makes you wonder about any omnipotent deity allowing people to believe they have an inside track. It is not really the fault of misguided, obsessed, or insane charismatic leaders with little conscience, craving to have their ego stroked. The fault lies with their gullible followers, compensating for their own emotional insecurities and overriding logic and morality. The answer is empathy, education, and tolerance in a society that does not frown upon freedom of religious and political expression, but where violence is simply not tolerated. Society is inordinately tolerant of some weirdo throwing rocks at police, because he or she thinks they are protesting for some dubious worthy cause. A non-negotiable one year jail sentence will soon stop that kind of behavior. No respect for police means you don't belong!

Feelings and Self-esteem
Too little self-esteem is the cause of too much,
and neither is helpful

Emotion is an amazing and fundamental human attribute. You may not realize it, but people lacking some compassionate emotion are generally disliked. Twisting that around, you can admire a person for intelligence and logic, but you'll only love someone with positive emotions. After millions of years of evolution we are experts at sensing emotions of people around us. That does not mean we always read it correctly because our own emotions distort the perception. Emotion expresses how we interpret our instincts, although it clearly overlaps with feelings. Feelings is a non-professional term for moods and awareness of events, impressions, intuitions, and hunches. It arises out of vague assessments of anything not yet subjected to logical scrutiny, after which it hardens as opinions.

Emotion is the mind (and body)'s reaction to feelings, sometimes the result of experiences and biases adopted early in life. Emotion can be a very significant obstacle, distorting views and hiding obvious but unpalatable truths. Emotional programming is largely defined in childhood but a sensitive child, raised by non-emotional parents or in an institution, can learn to control its emotions for better or worse. However, that is repression and it may increase anxieties later in life. Some people are far more emotional than others and very sensitive people are often condemned to a difficult life. Then again, they probably get more enjoyment out of good books, music, etc, and relationships and life in general.

'Feelings' are what guide and motivate us much of the time, subconsciously influencing priorities for what we say and do, and certainly not worth living without! Emotion is what gives us character and it is indispensable. Watching a typical TV news-report, about a plane crash for instance, will concentrate on how many people were killed. Then it suddenly switches to an interview with a friend of one of the victims, giving us a teary-eyed glimpse of what this person meant to his or her family and friends, and that transforms the entire report. It changes from factual to emotional, and it now required minimal imagination to grasp the disaster's human impact. This is far more important to most viewers than the high cost of the accident to some insurance consortium.

The most important semi-acquired component of anyone's personality has to be self-esteem, also formed in childhood. Children instinctively choose role models, selecting them as they grow, and these fortunate people have an obligation to instill a balanced sense of self-worth in the child. It is easily overdone, especially at the age of 3 to 5 when they must receive the message that the world does not revolve around them. Another critical time is the pre-teen years when the adult personality comes through. Then the message may be 'you are becoming a respected person, but only if you accept and treat others the same way! It is easy for children to develop insecurities due to intimidation, abuse, or an exaggeration of their deficiencies; and it handicaps them emotionally and socially in adulthood. Insecurity sometimes arises out of unsuccessful tries at independence, when parents obstruct with well-intended attempts to remain their child's best friend. Teenagers follow directions out of respect, eventually, wanting to look good in a parent's eyes without making it obvious. Excessive threatening may work temporarily but it destroys respect! If we are bothered by ugly and destructive behavior in society these days, self-esteem is the key!

Feelings, emotions, self-esteem and insecurity all combine to form a young adult's personality. The outward signs are an interesting mixture of assertiveness, humility, charisma, childishness, honesty, courage, and many other positive and negative characteristics that are quite recognizable. Personality can change somewhat over years but it is only adaptation, the foundation is genes and learned behavior during childhood. Young children need guidance and protection against bullying and without it they will arrange protection themselves by choosing suitable friends. And these may not be friends you would select! Bullying behaviour originates from territorial and domination instincts and can cause life-long feelings of insecurity in some children considered easy targets. It is traumatic when a carefree and protected child, used to easy friendships, suddenly must compete with or submit to other domineering boys, or girls, for popularity and status. Parents, teachers, or caretakers, not monitoring for signs of bullying are delinquent in a very fundamental responsibility.

Ambition and Napoleon
Survival demands ambition, but social
competition can be harmful

Ambition is a significant part of motivation, but it has many negative connotations. The desire to be more popular, important, bigger, richer, stronger, smarter, more attractive or powerful than others is in your genes because it is of major importance for survival. Most people are sensible enough to know that there is little advantage in making others dislike you by attempting to appear superior. But some have the intelligence, imagination, arrogance, acting skill, ruthlessness and lack of morals to try it anyway, and only deferring to people who can help them succeed. There are few better examples or better stories of unadulterated ambition then 'the Little Corporal' – Napoleon Bonaparte. The impact on his generation was comparable to that of Roman emperors conquering the ancient world. Few individuals ever achieved more and lost it all in such a hurry. Hardly anyone caused more suffering than he did, but it never bothered him much. He was the ultimate egomaniac, well equipped and primed to take advantage of the social foolishness and upheavals of his time.

He was nine years old in 1878, from a very poor semi-aristocratic family on the Mediterranean island of Corsica, when he was sent to France for schooling. All alone, and surrounded by rich French kids, very short and speaking poor French, he was ridiculed and bullied constantly. His fierce pride blocked all intimidation and made him a depressed loner, dreaming about revenge for the French invasion of Corsica nine years earlier. His intelligence, mature manners and dedication earned him a promotion to the Royal Military Academy in Paris and at sixteen he became a lieutenant in the French artillery. Although he impressed a lot of people, he was dissatisfied, unhappy, and still dreaming about glory for himself and freedom for Corsica. Unbridled ambition characterized him, and his meteoric career in the French army presented many opportunities to fulfill his dreams. His superiors all recognized his brilliance as a military strategist. In 1797, General Napoleon Bonaparte is 28 years old and a national hero after a very successful Italian campaign. Safely in Italy during the bloody revolution in France he takes his first steps towards political power by manipulating others to force a 'coup d'etat', removing the British supported royalist French government that had been established after the revolution.

He is successful because he manages to control everything his way (a control-freak he'd be called today). His soldiers love him for his good fortune and the plunder it provides, and his enemies are awed and fear him. He makes peace with foreign powers without consulting Paris and gets away with it because no one dares to challenge him. His ambition is exceeded only by his treachery towards anyone standing in his way and, as in war, he is totally indifferent to human suffering. His status and popularity in France grows tremendously (the art and other treasures plundered from Italy astound the entire nation) and in 1802 he is elected 'Consul for Life', with almost royal powers. A sentiment for the old monarchy helps him and in 1804 he simply crowns himself emperor. France was back where it was eleven years earlier, except Napoleon is a far more capable leader than King Louis XVI, who lost his head in 1793 at the height of revolutionary terror. For eight years after his coronation there were many threats, small wars, sea-battles and treaties with European powers, well documented in the history books. He was at odds with most of the monarchies in Europe but, after many military victories, he either conquered or dominated them all except Britain, which controlled the seas.

An interesting digression is the much revised set of comprehensive civil laws, the 'Napoleonic Code', that had already been debated for years to solidify French revolutionary principles. Many of the European countries adopted these laws, ordered by Napoleon, although they contained numerous undesirable features. Many countries today can trace their legal systems back to this code. It declared that all men are equal before the law (but not women!) and it ironically restored colonial slavery. Basic human rights had not been much of a reason for the Revolution; the majority of people were simply fed up with idle aristocrats and the Church.

There is no need to detail all the wars, mostly between Britain and France but other countries as well. We get the picture of how Napoleon became emperor and from there it could only go downhill of course. His mother was a very strong personality who retained considerable influence over him, but not enough to impact events. For the next while Napoleon managed to control things in his huge European empire, although he needed a large secret police force to do it. The beginning of the end came in 1812 when Napoleon broke his one-sided alliance with Czar Alexander and invaded Russia, setting fire to Moscow. The Russian winter thoroughly defeated the French and their mostly foreign army and they never recovered, never mind how much Napoleon tried. It is unnecessary to repeat a very well known story in history; other than to say that constant warfare and much bloodshed eventually drained France, while everyone else revolted. He abdicated and was banished (rather luxuriously) to the island of Elba. In less than a year his old friends and supporters organized an escape and he miraculously formed another army, drawing on French patriotism. Although his last campaign was brilliantly directed, Napoleon was narrowly defeated in 1815 by Wellington and the Prussian Blucher, who showed up at the last minute! Napoleon was again banished, this time to St Helena, and he died there in 1821, remaining a nightmare of shock, fear, and grief in Europe for years. So what was learned from this disaster caused by a single ambitious opportunist, responsible for the unnecessary death and injuries of multi-millions and the upheaval of families, towns, and nations all over Europe? Absolutely nothing, and that is the real lesson – about humans and their territorial, self-serving group instincts that can disregard the most horrific tragedies imaginable.

The blame was not with the little Corsican but with his subjects (the entire French nation) who supported him enthusiastically, riding his coat tails. Napoleon deserved to be a general, especially in that era, and he had more talent for it than anyone. However, when his obsession for power and control caused him to seek the dictatorship of France the hero worship and charisma should have worn off quickly, replaced by some objective thinking. The same sort of thing happened many times in history, to varying degrees. Hitler did it, and that disaster was even worse. It is not that supporters were unaware of the immorality (the rest of the world never tired telling them) but charisma, national pride, and blind enthusiasm simply won out over logic and reality. It is interesting that Hitler also ended up with the nickname 'the Little Corporal', although in his case it was not used by admirers.

There always were (and are) many Napoleons, although he was an extreme example. The ambitious and selfish, scheming manner is obvious, but lack of peer support usually stops any advancement. However, a small fraction are very intelligent and good actors, able to hide their negative motivations; they use charm at the right time and with the right people. They sneak through to higher positions, leaving behind a record of double-crosses and cheating. The key to their success is not that their ambition is unrecognized by superiors, but that the ruthless and scheming manner is useful to them. It is dangerous and self-defeating to openly oppose such people, but it can be satisfying to withhold support. Ambition is not always self-serving but it usually is. It is human nature to be motivated by material and social status, but there is a great moral difference between pursuing basic social security and an obsessive quest for status. Most of us would not sleep after violating the trust of a colleague, but such people don't even notice.

Choices and Relationships
We never seek the ordinary, only choosing it
when forced

It is a delusion that all things we do are important. They are of course, but only to us, and only at that moment. How secure we feel and how happy we are is always more important because that is the reason for doing it. We become obsessed with making a go of a new business because we expect it to increase happiness. We remind ourselves that there are many things that make us happy, such as time with family, feelings of joy by listening to music, going to movies, or reading a good book, holidays, and especially our relations with family and friends. Even Einstein pursued his discoveries for the happiness he anticipated by satisfying his curiosity and becoming famous, but not because it was immediately important to him. Of course, it turned out to be very important to society and no one grasped that better than he did. Family difficulties caused him unhappiness and it is likely that he regretted some early choices. It is judicious, albeit unattractive, to let long term benefits outweigh the far more exciting short term. However, there are no crystal balls for sale anywhere and emotional issues are impossible to judge. If it feels right, if it doesn't hurt anyone, and the opportunity presents itself, then go with it. There probably are good reasons for it; and Life leaves you behind if you worry to the point of inaction.

That may be a good opening to talk about the subject mentioned at the start of the book: Love! And I'm not singling out the obsessively romantic variety, because a semi-theoretical clarification of that has the same effect as getting drenched by a pail of cold water. The emotion of love makes acceptance of any such relationship an integral part of yourself, involving unconditional support that ranges from immediate family to your dog. There isn't a more common but accurate expression than 'Love makes the world go around'! We will defend who or what we love beyond reason, closing our eyes to what threatens such feelings. When our eyes are forced open, and love disappears, it brings on the most devastating and disheartening feelings. Love is meaningful and everything else is not, in the end! The opposite to love is hate, and both are deeply felt long term attitudes based on a mix of logical and emotional judgments; and our brain seems hardwired to accommodate them. Love instigates approach and hatred avoidance of people. But life is too short to even think about hate; we should be thankful for what it offers and enjoy it. Think of those banned to the Gulag for their entire life, cold and bleak, without comforts or food, no music, books, or hope.

Motivation is a simple word with a tremendously complex meaning. Even lower animals are motivated, but only by automatic instinct. Logical reasoning plays a major role higher up the intelligence ladder, to the point that most people have the impression they act logically all the time, although their opinion of others may be different. An interesting question is if it is possible to act entirely logical while paying attention only to socially positive motivations such as love, compassion, and general unselfishness, behavior we attribute to saints. Many individuals in history seem to have achieved this. But what exactly was their motivation? The vast majority of people are not like that, because evolution equipped them with selfish survival priorities. It is no use kidding ourselves, deep down we know that often we act selfishly although other times we do not, and it is destructive to feel guilty about it. We also like to believe that we only make logical choices and that we have a 'free will' which is every philosopher's favorite topic. Freedom from socially negative instincts should be an objective for a utopian society, but not for individuals because people will take advantage of it and ruin your life. We are not even close to a collective utopia and a 'social free will' usually means freedom to pursue collective self-interests, veiled by a social need to be compassionate.

It is ironic that the successful exercise of creativity is often accompanied by what we normally consider negative motivations. Examples are egocentric artists, ambitious leaders, greedy business people, obsessed inventors, etc; history is filled with them. You might assume that if saints dominated society our lives would be perfect; but in reality we'd be living in caves. Without selfish motivations everything would come to a standstill. Fortunately, motivations to create are much more complicated than that. A very big factor is the satisfaction derived from creating, even with relatively mundane tasks. Many people enjoy working due to the satisfaction it gives to do it just a little better each time. This book was written because it is satisfying to convert ideas into the right words, something conversation does not allow. Success (in society) and happiness (in life) are achievable only when everything is balanced to some extent. Unbalance may yield temporary victories but it will eventually crash around you. We inherited complex motivations to enhance survival chances and reproduction but it is not always easy to recognize it as originating from hidden predispositions. Even a mundane task like selling ice cream imparts a place in society; lose it, and life becomes precarious. The consequences are even more serious when mental or psychological illness overpowers such essential perception.

Although many actions are instigated by instinct, implementation and completion should be logical and that is entirely up to us. It is learned behaviour, although success often depends on character-traits. How we really feel about things is extremely difficult to manipulate, but with intelligence and practice we can regulate the outward responses, crucial in any social relationship. Social success is important and evolution has been careful to balance compassion with greed, love with aversion, etc. Winning is almost as important for survival as getting along, and it accounts for the abundance of human traits like ego, pride, greed, aggression, jealousy, passion, etc. Intellectually understanding our motivations and using the uniquely human ability to predict the response we may receive allows us to compromise and act wisely. Self-esteem is crucial and we must be secure in our ability to identify weakness and correct or compensate for it. Personal conflicts fascinate us; they are essential to all novels and movie plots.

Retirement makes you look back sometimes. It comes as a bit of a shock when many friends and family, taken for granted in your youth, are no longer there. Acquaintances from your working days and remembered with fondness have drifted away, except where friendship was deeper. Your new family has grown up around you and new friends are there also. So what is the problem? Well, there is that empty feeling when you allow yourself to think about it. It was not there early on, or even later when preoccupied with work. All were taken for granted, as if they would always be there. Much later you recognize the fallacy, but then they are gone! And whatever happened to those you knew casually, so important 20-60 years ago? Of course this feeling is not unusual, in fact it is normal, but it does explain why older people have a more difficult time saying good-bye.

Children and Motivation
Educated and motivated children are the key to the future

Probably you had enough of being lectured about things I'm not an expert in, although much of it is talking to myself. However, I can claim to know something about children, having been one myself and very fortunate to have two daughters and four grandchildren. Babies are born with an alarm system that enhances their welfare. Gradually they discover what works best and they use their developing logic and imagination to take maximum advantage of parental instincts. Eventually girls trick their parents into buying the latest fashions in clothing, important for their self-esteem, popularity, and social success, and therefore procreation. They become experts at social bonding and whatever else makes their life more secure. Boys practice their roles as family protectors and providers by mastery of sports, games, and technical things, in competition with others to prove themselves. All this still fascinates mature adults, sometimes conflicting with day to day necessities and maintenance of a healthy family-life. And you don't need to read this book to know all that.

What strikes me about children is that, apart from their preoccupation with friends and other non-consequential things, they have a great subconscious affection for people who are lucky enough to be a part of their lives. These selfish mini-people have strong emotional attachments, often unrecognized and therefore not capitalized on. You don't need to get excessively angry with them because their recognition of your disapproval is enough. Just let it simmer, it is respect for you that counts. Parents not respected and loved will never maintain discipline voluntarily. Most children are experts at making sincere friends, relatively untainted by the social motivations of more mature people. And sincerity and friendliness are not always closely related.

Logic and Rash decisions
Instincts will motivate but the implementation must be logical

The flaw in the criminal mind is that normal motivations are distorted. In many cases there is a critical lack of social compassion. Without the ability to cast yourself in someone else's shoes it makes sense to gain an advantage at the detriment of your victims, if it can be done without penalty. No sane person is entirely without compassion, nor is anyone without temptation to gain an advantage. However, logical thinking and judgment before acting is what counts. The average person will usually decide that socially negative thoughts are despicable, where the criminal mind will conclude that he or she can get away with it. Every decision, big or small, involves motivations with emotional and logical rationale. We are uniquely blessed with the intelligence and sensitivity to choose between selfish desires and social good will. It is maturity that polishes the skill of thinking before acting, making us feel good about ourselves by avoiding the rash decisions we'll regret later.

Emotion and imagination is what makes us 'human', more so than intelligence although some will dispute that. Without it we'd be robots, operating only on logic and preprogrammed instincts. We trust family and friends because of our emotions, not because they were verified as honest; we feel it and we sense the responses in others. The majority of individuals react similar in emotional situations, and it confirms our humanity and a promising future for our species. We should be thankful for our natural optimism, without being worried or depressed about the past; it is the future that matters!

CHAPTER II - <u>ORDER FROM CHAOS</u>

WHAT IS THE EVIDENCE?
Creativity and intelligence are strong directional filters for order in Nature

<u>Nature and Intelligence</u>
Intelligence is logical processing of senses,
memory and imagination
This subject ('Order from Chaos') should really have been the first chapter because it addresses some of the most fundamental issues imaginable! However, people important to me hinted they didn't like that! Oh well; it is unlikely anyone will ever discover how energy and the fluctuation that is our Universe originated, or how its instability invoked laws of interaction that created such unbelievable complexity and order. Some people suggest that the Quantum Theory of Physics explains it all, but that is exaggeration. Although we assume that the word 'Vacuum' means 'nothing', Quantum Theory holds that in this Vacuum (where there is no matter) the density of energy is so great that (virtual) particles can pop in and out of existence like bubbles in soda water and temporarily create order out of chaos. Some experts believe that our Universe is an enormous version of this, where matter-particles are discontinuous Vacuum-fluctuations in an infinite ocean of energy. 'Chance' is their explanation, paid for by a reduction of order elsewhere. For instance, a chance merger of three nuclei at nearly the same instant can create an avalanche of order within a hot chaotic gas, and reduce local temperature.

Anyway, my friendly advisors were right, that would be too deep to start a book with. Also, the nature of energy and its origin(s) are questions so fundamental that it is inappropriate to speculate; and don't expect any answers in the near future, or ever. There have been proposals suggesting how it sustains our physical (particle) world and that subject will be covered in Chapter III called 'Energy and Substance'. The origin is postulated to be entirely unknowable. Much easier to understand is that nature has little trouble in creating order out of chaos and does it all the time. In its most complex form we call it 'Life'. The source of life is also an issue we ought to stay clear of, since logic and religion are opposites, but it is too interesting. The advent of intelligence is one of the best and most fascinating examples of order. Not that intelligence as such is orderly; it has my vote as the most unpredictable process in the Universe, because it uses instinct and emotion to guide it, in addition to logic. However, combined with imagination it is capable of identifying sources of chaos and the creation of order by intervention.

As such, the rise of human intelligence, combined with imagination and artificial intelligence, is unique in Nature. Nothing else has the potential for unlimited growth without running into some natural barrier to oppose it, such as a shortage of food, space, disease, predators, etc. The only evolutionary limit on intelligence may be social instincts; they will always be more important for success in procreation. Not too many eggheads in history were known as eminent lovers!

<u>Thermodynamics and Energy</u>
Gravitation bestowed order on the dispersion of
'Big Bang' energy

The consensus is that order thrives on Earth because we have an 'open' thermodynamic system, with copious energy supplied by the sun. In the 18th and 19th centuries it was believed that order can only decrease, and that increasing order was therefore evidence of supernatural influence. Today we know that the 2nd law of thermodynamics (overall disorder can only increase) means that when nature creates order somewhere it is at the expense of more disorder elsewhere. It is not generally recognized that the instability that created the Universe (the 'Big Bang') must have been responsible for any order we see today.

The sub-atomic building blocks of our physical world possess unique attributes called 'Forces of Nature'. And without the self-gravitation of huge gas-clouds, collapsing into stars hot enough to fuse atoms, there would be nothing at all! Electro-magnetic and sub-nuclear forces are equally necessary to stop gravity from converting the energy stored in elementary particles into pure radiation or whatever other form Vacuum-energy has. This may probably happen anyway, at the end of the day, and it suggests how our Universe got started. Science now generally accepts Darwin's Theory of Evolution and also mutations as the driving forces behind biological order, although religious conservatives dispute this. Experts are confounded by the circumstances surrounding the emergence of self-replicating molecules,

56

the forerunners of life on Earth, but speculative theories abound as to how chance made that inevitable.

Civilization and Society's Future
Only our genes and civilizations provide any
human permanence

Another step change (mutation) in the evolution of human intelligence on Earth is unlikely; being smarter no longer favors survival as much as it used to. Our global society is too big, open, and successful and we may end up like the horseshoe crab, which did not change much for 400 million years. Of course, intelligence is a factor in selecting a mate and a gradual increase is therefore expected, although bigger brains requires a bigger head, and that is problematic. There will always be individuals with significantly higher general intelligence than average, but their impact on the average population will be minimal unless they all emigrated to some sparsely populated area with harsh living conditions. If evolution changes humanity in the future it may be our social instincts, because that now applies pressure on natural selection. However, there are a few questions: Are we at the end of an evolutionary experiment and the beginning of a new one? Also, can selective social evolution develop greater social order in a society of billions of humans? Nature almost seems to respond with 'OK, you have arrived; now play your part and organize yourselves'.

Where this will lead is uncertain, although a similar problem must have faced the ancestors of bees and ants a very long time ago. Thanks to the media, who make money reporting conflicts and violence caused by vested interests and prejudices everywhere, we might conclude that our civilization is headed towards oblivion. But who knows, we may get lucky! Seriously, we all know that our negative impact on nature is increasing exponentially. If we want a more positive outcome (since we can't go back) we must reinvent human society, voluntary or forced by progressively severe catastrophes. Of interest is what human society may be like a thousand or even a hundred thousand years from now. This invites the usual comments: 'Why should I care, I won't be there!' Staying away from religious arguments, I think our society's future is far more interesting than who wins soccer's World Cup. In fact, the World Cup and other such events seem to bring out territorial aggression and a war-like thrill of conquest in people, and it bothers me.

Social order from Religion
Organized religions were often responsible for
social order

Although arguments for or against supernatural influence are pointless because evidence either way is conspicuously absent, it is still possible to draw at least one logical conclusion. Based on direct evidence alone there is little doubt that external control of our daily life is (predominantly?) absent. However, the human propensity for the supernatural and the optimism it instills is so strong that it must be a fundamental part of us, and this is where we should look for evidence! Religion relies on it - our innate expectation that there must be something, before and after death. There is no question that religion has been a catalyst, causing more order than chaos in our societies. It is ironic that even religious wars seem to increase order, after the opposition is defeated. The expectations of some were usually achieved through the deaths of many. Of course, the word 'religion' does not necessarily mean acceptance of God(s); in many cases it implies the acceptance of the attitudes, disciplines, superstitions, or life-philosophies dictated by an elite group, propped up by a large but incognizant membership. It is not surprising that religions are strong on tradition, customs, and celebrations; it makes people forget and feel good ('Comfort and Joy' according to the old Christmas song).

Religious leaders of all faiths tend to assume that children grow up and will accept faith by indoctrination. After all, they did, and the children's parents did, and their religions make sense with rational messages of hope and guidance on how to conduct your life. It is understandable, because certitude that you are not alone is a huge comfort to an individual. These leaders also assume it reasonable that adult outsiders would want to join and convert to their religion. Such an assumption bewilders me, since you need to suppress all notions of reason or logic to convince yourself.

For instance, you must put aside the infinite evil you read about in history books, when religious leaders initiated witch-hunts during the 'inquisition' in the 15th and 16th centuries, horribly killing a hundred thousand mostly single old women in Europe alone. Villagers who suffered a bad harvest or other disasters were often the accusers. These poor women were then tortured and forced to admit guilt, sometimes for having sex with the devil, and none were ever found innocent. We can blame all of that on ignorance or superstition and a craving for depraved excitement, but we cannot ignore the fact that all these 'witches' were convicted by courts staffed or appointed by religious leaders.

Why worry about that in this modern era? Well, many religions still give credence to angels, the devil, and demons; and witches are not that far off. It always seemed strange to me that such stories were openly accessible, even part of folklore, but that any mention of sexual activity was taboo when I was a child. Witch-hunts are picked as an example of cruel and irrational behavior but there are others, and not all of it is ancient history! Religions can be very positive and often an essential force for social progress, but they are human institutions with human weaknesses and the fall-out from such weakness must never be swept under the rug, although it usually is. There are no infallible humans alive today!

<u>An Open-ended Future</u>
Our future is indefinite, guided by chance, logic,
and instinct

We see no pre-determined plan; life at all levels appears to require no intervention and progresses chaotically all by itself. If there were an inflexible plan it would never have allowed us the hypothetical power to wipe out all life on Earth at this dangerous stage! An open-ended future may be possible, based on chance, and advancing step by step to create experimental entities such as humans to write the next page, and so on. Creation of humans in God's image is a concept that was clearly conceived to claim the opposite: God must be like us, except perfect. Why would an omnipotent Supernatural, capable of reaching any destination without experimenting, bother with such a long road?

This thought allows a look at supernatural questions from another angle: perhaps 'nothing' is impossible and whatever exists is an intrinsic consequence of instability and evolves along with us, toward an open-ended future. That leaves a lot of room for intervention, or guidance. From ancient times on all but atheists have asked: 'What is expected of us?' A possible answer is that very little is expected of individuals, but much of human society. Foremost, we must quiet our egos and accept that we are likely not unique, that experimental evolution goes on everywhere in the cosmos. And who knows, it is possible that atheists are right; there may not be anything, and then it is all up to us! Who wants to play god?

Growth of global order for humankind should come primarily from social evolution, continuing to develop from our tumultuous past.

We should hope that it will be with more compassion and less bloodshed. We underestimate the enormous contributions made by cultural heritages from all parts of the world – our racial diversity, religions, attitudes, agriculture, art, literature, history, etc. Social evolution is every bit as complex as biological evolution and it also will turn into a major science in the future and be of primary importance to future generations! However, we desperately need changes now! Aid for disadvantaged children everywhere must be improved, specifically medical care, food, and education. The richer countries can easily afford that, provided it is organized without the usual political misdirection. The present situation in some areas is detestable. Of course, there is a fear that the standard of living in developed countries will suffer, but that is doubtful. Close ties with under-developed countries provides unforeseen benefits for both sides.

The point is that a functional global society is not just desirable but imperative and it starts with educating (and feeding) all the children. Failure to attempt at least that much is socially immoral. Feeding and educating your own children with little regard for others may be natural but it will lead to extinction! The United Nations could easily coordinate a global plan supported by all developed nations to establish schools in the right locations, and administer supply and staffing. Leaving it up to under-developed countries to make-do with a few hand-outs accomplishes very little. Education unlocks the promise of intelligence and imagination and is the road to the future.

The buzzword in the nineties was "global economy", but its shine is now gone. The rich were too eager to exploit cheap labor and become richer. Capitalistic shenanigans caused a major global economic depression and protectionism tightened many borders again. However, the ball is still rolling and it won't stop; countries like China now have the means to finance their own economies and will continue to flood the world with lower priced goods, at least for the next while. Without a fair global economy many countries will be excluded and we all will suffer the consequences!

ORDER BY COINCIDENCE
Order out of chaos is created by balanced opposites, with one defining the other

Accidental Existence and Laws
Science believes the laws of Nature are
responsible for order

People who lean towards an atheistic point of view, that there never was any supernatural influence, must also accept that our Universe accidentally received some outrageously improbable combinations of laws and constants of nature, allowing matter and life to develop. One possibility is that, out of a multitude of different regions or Universes, we are in one that has the right combinations. Or, alternately, that the properties of Vacuum-energy and its fluctuations are such that matter and life will eventually form under just about any condition, in near-infinite variety. Logic convinces atheists that supernatural influence is impossible, although the bleak consequences of a purely accidental existence is difficult to deal with.

The primordial Vacuum (prior to the Big Bang) contained near perfect order or disorder, and that could be the beginning and end of a 100% efficient cycle. Our existence proves that this cycle has the potential to form complex biological life, pre-programmed and leading to a conscious logical intellect capable of self-manipulating physical and biological forms. The Vacuum was orchestrated into specific order, directed by fundamental laws of nature. Biological evolution is supposedly controlled by environmental adaptation and proceeds randomly towards greater specialization, but in reality it evolves towards greater sophistication and complexity. Those of us who assume there was (is) no guidance must accept there are natural laws beyond Darwin's to force this tendency. Any purpose, other than survival and adaptation, is detested in evolution-theory but why does Nature not backtrack, from pterosaur to eagles and back again to penguins, or whatever has the best chance of survival? Instead, it only advances, from lemurs to apes to humans. That suggest there is more than just environmental adaptation and survival at work.

Biological evolution and order resulted from Nature's magnificent creation of DNA and its progressive changes by mutations and natural selection. Unsuitable forms become extinct and competing mutations take over. We still carry the DNA of all our predecessor species, in addition to the specific successful mutations of Homo Sapiens. Investigation of Nature's shrouded origins invariably forces searches for additional underlying causes. Intelligence may be an exception because some mammal species are far more intelligent than they need to be; and we are a prime example. Another example is dolphins; sharks have the same diet as dolphins but they don't compare in intelligence, although both species are very successful! There was no real need to replace the Neanderthals, they were successful, but the brains of all social animals seem spontaneously receptive to mutations that advance logical thinking. Intelligence is clearly helpful to some species and not limited to mammals. Birds are reptilians and some, parrots and crows for instance, are quite intelligent. Intelligence is the key to the Universe; it facilitates action, not only natural reaction. Our imagination and mastery of languages and mathematics, combined with unlimited A.I. memory and processing, renders a creative ability suggesting a metaphysical link sometimes. Although easily overlooked or disclaimed, the enigma should be obvious!

The creation of carbon-based life-forms demands a very unique combination of laws of nature that can only be called astonishing, if a coincidence. Most scientists believe that life arose on Earth spontaneously because conditions here happened to be exactly right; we were simply lucky there was this unique place. We can only exist where conditions for our biology are compatible and this is their explanation. The scientific name for it is the 'Anthropic Principle'. However, you are probably scratching your head right now, like me. From all cosmological test-data it seems that the laws of nature are more or less the same everywhere, even to the edge of the Universe. If that is correct then there must be many other places with biological life since our sun is one of 10^{22} stars (one with 22 zeroes behind it); and why should Earth be unique? There could be an infinite number, if there are also an infinite number of Universes as some people suggest. Infinity is a concept we cannot handle and in this case it also means that there must be an infinite number of planets with intelligent beings like humans walking around. Science's Anthropic Principle is not concerned with any of that, it simply explains why we have been a success story here on Earth.

It does not explain how life developed here spontaneously (and was never extinguished!), knowing that the probability of finding the necessary combination of conditions anywhere is nearly infinitely small.

However, scientists hang their hat on the fact that this probability is not zero, since we know we exist! Given the enormous number of chemical opportunities and available time, they are convinced that the formation of life was inevitable, despite its improbability. However, in many ways the Anthropic Principle resembles a scientific cop-out because it provides a conveniently ambiguous defense against religious arguments that life is a miracle.

Hoyle and the Anthropic Principle
Chance formation of life's atomic structures is beyond belief

If you are undecided about what Religion or Science are suggesting, then you really ought to acquaint yourself with the many improbable natural coincidences Science uncovered over the last 100 years. In some cases it even changed the discoverer's mind about the role of chance in the creation of our Universe and of biological life, although not always in the same direction. A good example is Sir Frederick Hoyle (1915-2001), a controversial British astronomer who discovered that carbon, which our life form is based on, is manufactured inside stars only due to an implausible coincidence.

All early stars began as gravitating (clumping) clouds of gas, consisting of hydrogen and helium. When enough gas collected, self-gravitation compressed the denser regions and escalated pressure and temperature until nuclear fusion lit up, converting more hydrogen into helium and creating a star. The energy of four separate hydrogen nucleii is greater than for one helium nucleus and surplus energy is therefore released as radiation (photons), further escalating the temperature inside the star. Fusion merges four hydrogen nuclei (protons) into helium, two of which convert into neutrons. The necessary energy for this is easily supplied by the enormous kinetic energy (particle velocity) inside the hot star.

Although fusion of helium nuclei into heavier elements produces very stable elements, such as carbon and oxygen, this is not the case for beryllium that is in-between, consisting of two helium nuclei. It is very unstable and can only exist for an extremely short time and therefore it is not present in stars. This should have prevented the formation of carbon and oxygen, with three and four helium nuclei resp. This is where Fred Hoyle enters the picture, concluding that since we are here these elements must have formed somehow and there must be a star fusion process that allows it. He predicted that it occurs during the very brief time (10^{-17} sec.) that a beryllium nucleus can exist and he calculated that, based on the structure and predicted temperature of a star's core, a specific quantum energy level (resonance) of the carbon nucleus must exist. This is analogous to two ships that can tie together only when both have the same velocity vector. Experimental results proved him right, against all scientific common sense at the time, and it opened the door for calculation of the fusion processes that created all other elements inside stars.

Hoyle's prediction was remarkable, and not only because it was the first (and only one so far) based on the 'Anthropic Principle' (we are here, so it must be true). It may be more than just remarkable that such a resonance would exist at that precise level; a slight difference and we could not be here and neither could the Universe! Of course, many people see this as evidence of supernatural influence. Atheists consider this nonsense and will point out that different laws of nature could produce other life forms or Universes and that ours is therefore just a random choice. I am dubious about this because no one has been able to suggest a Universe slightly different from ours with the potential of even supporting matter and life, never mind spontaneous formation.

Hoyle himself was thoroughly shaken by his discovery and rejected his atheistic beliefs. In his own words: --*"Would you not say to yourself, some super-calculating intellect must have designed the properties of the carbon atom, otherwise the chance of my finding such an atom through the blind forces of nature would be utterly minuscule. Of course you would . . . A common sense interpretation of the facts suggests that a super-intellect has monkeyed with physics, as well as with chemistry and biology, and that there are no blind forces worth speaking about in nature. The numbers one calculates from the facts seem to me so overwhelming as to put this conclusion almost beyond question"* (Hoyle: 'The Universe: Past and Present Reflections.' 1981).

<u>Creationists and the Big Bang</u>
Paleontological evidence disproves all claims by
creationists

Although creationists used Hoyle's supernatural speculations to support their views, such assertions are disproved by irrefutable paleontological evidence that all biological life on Earth evolved by mutation and natural selection. Impenetrable initial conditions and the properties of Vacuum-energy must have combined to mold the laws of nature, compelling the Universe to develop in a direction that allowed formation of biological life and intelligence, with or without help. It seems that the path exists and the eventual end result is inevitable, because chance allows it and time is endless; Nature only has to keep on trying. It seems the real mystery lies in initiations, of Vacuum-energy and of biological life. All that is conjectural, as it will always be! We can think of life as a game of 'Scrabble'; you're given a bunch of random letters and sometimes it makes very good words, but only if you can come up with it. The potential is there, but the implementation is up to you; and if you don't find it, you'll never know what could have been!

We shouldn't leave Fred Hoyle yet. Repeating the importance and possible meaning of his discovery: - Carbon is formed by fusing helium and beryllium, but beryllium is not present in any star! Had beryllium been slightly more stable, the whole Universe would be nothing but carbon! And if the carbon resonance had been slightly lower or the resonance for oxygen slightly higher, there would have been no carbon! This tiny window of opportunity for our life form has caused many famous physicists to suggest that the coincidence is inexplicable (a 'put-up job', according to Hoyle). Second, Sir Fred became an outspoken critic of the 'Big Bang' theory (which he coined in a sarcastic moment), advocating the 'steady-state universe' theory instead. However, he and his collaborators were proved wrong after the accidental discovery of cosmic microwave background radiation in the 1960's.

It is also interesting that Hoyle promoted 'panspermia', the concept that life began somewhere else in the universe and was transported here. He was not the only one; there was a lot of interest when it was discovered that carbon-based molecules similar to those in living things drift around in space. Unfortunately, Hoyle's hobby was writing Science Fiction and he did not have the personality to tone down controversial opinions, and it reduced his otherwise considerable credibility. However, the 'Big Bang' issue is not completely settled yet because there is no credible concept of what came before! Also, many scientists today are convinced that formation of the first replicator by chance is highly improbable. Quoting Hoyle again: - *"The notion that not only the biopolymer but the operating program of a living cell could be arrived at by chance in a primordial organic soup here on the Earth is evidently nonsense of a high order."* He did not revolutionize the world, and he was definitely wrong on some things but he was not afraid to stick his neck out (very unusual for a prominent scientist), and Fred Hoyle is therefore admired and respected by many people.

Cosmological Coincidences
The Universe is a battleground between gravity and electro-magnetism

There are even more obscure coincidences, discovered and still debated by many cosmologists. One relates to heavy elements, also essential for life on Earth and produced only in the unimaginable density that occurs during the Supernova implosion/explosion cycle of a massive star. In large stars the fusion process stops at iron and they end up with an iron core. However, that is not really the end because the lack of fusion and subsequent slow cooling causes unstoppable escalation of gravitation pressure, slowly forcing protons and electrons to combine into single particles as neutrons, occupying only a fraction of the original volume. At a predictable point the iron core collapses gravitationally in less than a second and this releases so much kinetic energy (temperature) that, during the rebound, nuclear fusion spawns all conceivable combinations of stable and unstable elements.

During the inevitable (Supernova) explosive rebound this chemical bonanza is distributed throughout space, propelled by an enormous number of neutrinos, released when protons and electrons merge. The neutrinos carry away 10% of the star's rest energy but they (and other mechanisms) are also responsible for blasting the star's outer material into space. The coincidence here is that this process is ultra-critical, with a little too much or too little being equally disastrous. Instead of seeding the Universe with essential elements for life, it would cause everything to fall back into the star. What is left after a Supernova explosion is a 'Neutron star', consisting of the densest material still detectable by photon emissions. A soup-spoon full has the equivalent mass of a cubic kilometer of water, if you can imagine that!

Another major coincidence is the uniformity and homogeneity of the Universe. It must have been entirely uniform before the 'Big Bang' even started because there was no chance of smoothing it out later. Its rate of expansion was unbelievably critical; any deviation smaller than what you can imagine would have caused it to collapse long ago or prevented any chance of star formation from unchecked dispersal of radiation. In either case, the Earth in all its glory would not exist!

There are many other coincidences. For instance, life on Earth is protected from lethal radiation by the Sun and outer space by its unique magnetic field. Without such protection we could not survive deadly genetic mutations, but the present low rate is just right to support evolution by natural selection. Also, the Sun's radiation will remain stable for another 6 billion years, due to negative feedback maintaining its fusion-rate at a constant level. The Sun expands when its fusion-rate increases, decreasing the core temperature and thereby its rate of fusion, a very lucky coincidence for people on Earth. Not relevant actually, but a strange coincidence anyway, is that during a full solar eclipse the Moon exactly covers the Sun. If you lived 5000 years ago, would that spectacle alone not convince you that God(s) were responsible? The Sun's diameter is 400 times larger than the moon but it is also 400 times farther away.

Without any doubt the biggest coincidence of all has to be the formation of our physical world out of unstable Vacuum energy, which we cannot even sense and unaware of until recently. Scientists speculate (based on convincing circumstantial evidence) that all the stable building blocks of matter were created during the so-called 'Big Bang'. However, the processes that followed are so intricate and improbable that we must wonder how it all fell into place. It has been compared to a playful monkey arranging a million random letters of the alphabet, and ending up with the Bible. Many people, well-known scientists among them, have suggested that the Universe expected intelligent beings like us, and prepared the way. That may be a little presumptuous because the rise of intelligence makes perfect sense from an evolutionary point of view; it just happened to be us and not dolphins or chimpanzees, for instance.

Science tells us that Vacuum-energy forged a small part of itself into immense quantities of elementary particles during the Big Bang. They interact through four distinct forces, essential for stable particle-structures. Such structures were fused inside stars by enormous gravity and high temperature into larger structures that can chemically combine in millions of ways into anything a chemist might wish for. The accidental (?) formation of self-replicating molecular structures then allowed biological evolution to produce chemists and every other living thing. Atheists listened to scientists trained to believe in cause and effect, claiming that it is all caused by the 'Anthropic Principle' – (i.e. all these unlikely coincidences must have accidentally happened, because we are here!). That is philosophy and, like all philosophy, it proves or explains nothing. There are probably many possible explanations, but it is doubtful that Science will ever get anywhere near the bottom of it all.

ORIGINS OF BIOLOGICAL ORDER

Though I have no doubt that the origin of life was not in fact a miracle, I do believe that we live in a bio-friendly universe of a stunningly ingenious character (Paul C.W. Davies)

Origins of Life and Evolution
Circumstantial evidence suggests there is bias in Nature towards Life

Life Science is the study of the structure and function of self-replicating living organisms. Behind the incredible and intimidating science you see on television or in magazines are a few unanswerable questions -- what is life, and how did it get started? Most of Science holds that evolution's path towards a human existence was chemically inevitable, given the laws of nature, the passing of time and opportunity (if you buy enough lottery tickets you win eventually). Convinced of a basic supernatural initiative, many non-scientific people listen to a minority of scientists with religious beliefs, claiming that the extraordinary combination of coincidences (laws of nature, the formation of complex life, etc.) is impossible to have occurred only by chance. Opposing this are atheists who believe that the concept of a God creating life, or merely starting it, is even less likely. As mentioned before, the laws of nature appear the same everywhere in the Universe although this is questionable since we are like ants living on a tiny coral island, incapable of seeing anything non-local. Some cosmologists propose there are an infinite number of Universes, increasing the odds of life beginning somewhere to 100%. Of course, infinity also increases the chance that a super-cow can eventually jump over the moon to 100%.

Atheism, Science, and Purpose
Human 'free-will' seeks objectives, hence Nature has Purpose

Atheism is the unofficial religion of Science, and few prominent scientists will admit to believing in any supernatural influence over biological life. They are also silent about a possible supernatural beginning, although there is some sympathy for that (many of them are really agnostics but not interested in arguing about it). Their real foe is the concept of "progress" in evolution because it suggests "design" in nature. For an agnostic that is a little stifling, with nowhere to go but to accept a lack of purpose and direction in nature. That may be the way it is, but it remains an open question until proven with hard evidence! Science suggests evolution is like a fictional game: starting with two dice, you add another after throwing two sixes, etc, eventually ending up with so many dice that the complexity with each throw holds every possible sub-combination imaginable. Scientists are right; that is not progress but increased complexity (which is how they view evolution). But they forget to ask where the dice come from! The real mystery is the Vacuum and its laws that, against odds as large as our Universe, created an absolutely mind-boggling planet of wonder with intelligent biological life as its star-feature.

That at our stage of evolution advanced intelligence appeared was probably inevitable; and Darwin's theory is responsible. However, humans really are special; our ability to logically observe and analyze ourselves and then imagine situations or circumstances in our past or the future, and remember it, is miraculous. Control of human life was seized by the alliance of intelligence, imagination, speech, and instinct and that accounts for the jeopardous success of our species. Assuming that this enormous step was governed by chance is hard to believe, but then, possibly not. It may be that nature was ready for it and simply exploded. Without evidence I insist on retaining my emotional attachment to some initial reason or purpose behind it all, and anticipating hidden laws of nature (beyond Darwin). Besides, Science must be wrong about a lack of purpose, direction, or motivation in nature. We evolved from and are full participants in Earth's ecology, and our motivations (instinct and intellect) control everything we do! We are part of Nature and motivated by many purposes, therefore Nature has Purpose, at least in our case! Our stampede towards constructive complexity in society, with its unimaginable consequences, strongly suggests hidden objectives. With or without a conscious (or oblivious) supernatural influence there could be hidden directions in nature and in human life and society!

<u>Organisms and Definition of life</u>
The mystery of life resides in the information-content of DNA

Order in living organisms appears to be self-organized and spontaneous. This self-organization relies on natural selection and mutations in vaguely understood ways, yielding the profuse magnificence of our biosphere. Life can be defined as a vast sequence of complex chemical processes that form replicating or self-reproducing organisms. But that does not really go to the heart of the matter, because these chemical processes are controlled by an astonishing and identical information system (DNA) that is present in every cell (the basic unit of life) of a living organism.

At least in theory we may be capable of creating robotic life in the very distant future. They may be technically intelligent devices programmed in such detail that they reproduce and reprogram themselves, including the processing of necessary data and materials. Would that be life also? If not, our definition of life is incomplete and there is a difference between 'artificial' and 'biological' life. Life is not contained in any single molecule; it results from the coincidental joining of large numbers of very complex and unique molecules or groups of molecules, acting cooperatively to create and conduct life. An intriguing certainty is that these joined complex molecules are essential and could not possibly have existed on Earth before the first replicator formed, although their constituents did.

Sequencing the DNA of human genes has been a precursor to investigating the folding or shape and structure of human protein, a matter of equal importance and with only a few percent accomplished so far. Genetic engineering has the potential to conquer cancer, grow new blood vessels in the heart, block blood vessels in tumors, create new organs from stem-cells, and perhaps reset the genetic coding that causes aging, among many other possibilities.

How will evolution change humans? It seems probable that the individual's concern for society as a whole will increase in the future. Anyone low on social instincts is bound to be less successful, in life and reproduction. Today, human evolution is at a crossroad between individual and social motivations, exemplified by such diverse activities as robbery and social assistance to the needy. It is fascinating to speculate how this will develop in the distant future, not knowing if we are being led and hoping that our civilization will last long enough.

DNA and Self-Reproduction
All biological life relies on sequencing and
folding of DNA

A common scientific opinion is that the spontaneous formation of simple natural replicators is expected (or inevitable) given the available time for accidental opportunity, the conditions on a young Earth, and the building blocks (amino acids) known to exist at the time. One theory is that these amino acids combined and formed into early self-reproducing RNA (ribonucleic acid), eventually evolving into the similar but incredibly complex codes (molecular sequences) contained within the DNA (deoxyribonucleic acid) of today's life-forms.

DNA duplicates itself by first unwinding its two loosely tied helical strands with the help of specific enzymes (the strands are far stronger than the ties between them). It then simply relies on chance and a generous diet of four specific organic compounds (nucleotides) floating around in each cell. Two combinations of the four bases are possible for each rung on the DNA-ladder (and their mirror images) but, by a very convoluted chemical process, copying of specific sections of DNA determines the types of proteins produced for assembling any specific organism.

Each individual has the same DNA sequence in all its cells, and physical variance between individuals and species is caused by DNA differences. James Watson, Francis Crick, and several collaborators discovered the structure of DNA in 1953. It is fascinating to speculate how human traits such as compassion, morality, and creativity are codified within the sequencing of DNA and also how much of it is acquired and therefore stored in memory. It is obvious that life at higher levels needs both biological form and life-experiences (i.e. 'hardware and software'). Compassion is a gift that aids social interaction and resides even in animals, although some people appear to be rather immune to it.

Don't get the impression that I know what I'm talking about, but apparently every child randomly receives half of its DNA in 23 chromosomes from each parent, for a total of about 3.1 billion letters of code (nucleotide combinations). There appear to be roughly 25000 protein coding genes in the human genome, which utilize only 1.5 % of the available DNA-sequences in our cells. The rest is excess baggage, left over from primeval evolutionary steps, and mostly useless. The quantity of genes in the genome of an organism seems irrelevant because we have roughly the same number as a worm or a small plant. It indicates that in basic form we are not unique in Nature, although we discretely claim evolution selected higher quality genes for us, like aristocrats claimed in previous centuries.

Natural selection and Mutations
Nature does not design life, it only rejects
inferior versions

Mutations followed by natural selection spur the evolution of life on Earth. Mutations are random variations (copying errors) of DNA sequences, made permanent only by preferential reproduction (natural selection) when it improves an organism's chances for survival. Natural selection is automatic; organisms physically or instinctually unsuited for their environment die early and lose the competition to reproduce their DNA. This is a good example of self-organization, how nature creates order out of apparent chaos. It selects not only obvious advantages such as strength, bigger teeth, or faster running, but often nebulous things such as the mane of a male lion, or a man's beard. This presumably makes them more threatening and successful in defending themselves (and more appealing to the females of the species). Although natural selection of beneficial mutations clearly increases order, it depends on reproductive replication and therefore cannot be responsible for the so-called 'origin of Life'.

Mutation of DNA molecules in living cells is normal; and in the vast majority of cases it is destructive, or has little effect and dies out. Very infrequently, under long-term environmental stress, specific variants of an organism have some small advantage. If it manages to expand to a critical number the odds are that in the long term this variant will take hold in the entire species. In this manner organisms slowly adapt to their environment, by changing the information contained within their DNA. This has implications for humans because social stress now dominates our lives and some mutations may eventually be selected and change the social instincts of our species.

Biased Universe and Origins of Life
Life started somewhere and somehow, with
organization of information

The objective of this book is to point out areas of uncertainty in our knowledge, but not to provide expert information. Knowing even less about biology than many other things, included in here (with his kind permission) are a few pertinent and compatible quotations from a book by physicist Paul C.W. Davies ('THE 5th MIRACLE'– The Search for the Origin and Meaning of Life, Simon & Schuster, copyright 1999 by Orion Productions). Davies has written many interesting science books, among them "The Mind of God" and "Are

We Alone?". He was awarded the 1995 Templeton Prize for his work on the deeper significance of Science. Although a physicist, Davies spent two years researching life's origin, and he concludes:

- *"I am now of the opinion that there remains a huge gulf in our understanding. To be sure, we have a good idea of the where and the when of life's origin, but we are a very long way from comprehending the how".*

To put his use of the word 'miracle' in the title of his book in the right perspective, the following is quoted, also out of the preface:

- *"Though I have no doubt that the origin of life was not in fact a miracle, I do believe that we live in a bio-friendly universe of a stunningly ingenious character".*

This last quote probably represents the opinion of a significant number of scientifically informed people. It leaves the door open to many possibilities. The book is well balanced and it carefully avoids upsetting any faction too much. It presents the available data and draws logical conclusions, although logic is a word that should not be associated with the uncertain origin of life on Earth. This book is certainly recommended for anyone interested in the subject. The following Davies quotes are listed without further commentary, but please remember that such quotes are easily taken out of context and only mentioned here because they make very good sense to me:

- *Life's distinctiveness lies not in the chemistry as such but in its informational properties.*
- *A living organism is a complex information processing system.*
- *The ultimate problem of biogenesis is where all the biological information came from.*
- *We are missing something very fundamental – A fully satisfactory theory of the origin of life demands some radically new ideas.*
- *It seems that life must have begun as a ramshackle process and became refined and streamlined over time.*
- *Successful mutations are those that are better adapted to their environment, and it is therefore the environment that selects the information that ends up in DNA.*

Davies also mentions three current scientific theories for biogenesis: 'chance' (Darwin's 'warm little pond'), 'panspermia' (life imported from space), and 'geothermal activity' inside the Earth, or Mars, or both. He identifies how nature found a way for living organisms to selectively incorporate informational changes in DNA, to become more compatible with the environment. He addresses the question of how the first biological replicator formed (biogenesis) but supports neither chance nor a miracle, suggesting instead that there may be a bias in nature towards the formation of life.

ORDER FROM KNOWLEDGE
The spread of collective knowledge enhances social homogeneity

Language and Mathematics
Language is our secret and we are doomed
without it

The application of intelligence to information by searching and rearranging it into logical patterns is a magical filter for creating order out of what is otherwise chaos. Human society cannot function as such without communication and language. Our brains are hardwired to accept and learn any version of verbal or symbolic language and to apply links between combinations of words or symbols and images by painting virtual pictures in our mind. We know that a baby from English parents can grow up in China and accept Chinese as its mother tongue without difficulty; our brains are all the same!

It is difficult to rationalize anything without the ability to describe concepts in common language because language attaches labels that provide meaning. We think 'tree' and subconsciously ask 'what kind of tree'? From the answer we imagine its shape, color, etc. We create order by visualizing and selecting imaginary arrangements based on memorized knowledge or knowing where to find that information, all by using combinations of language labels (words). Encyclopedia, textbooks, and the Internet are useful because we follow well-marked label-trails until reaching a conjunct end-point where everything makes sense and order is created where it did not exist before (in our mind).

Mathematics is the universal language of logic, with images and symbols that convey valuations, equations, and processes with precise rules, unlike ordinary grammar that can often have more than one meaning. All knowledge is encoded linguistically but specialized knowledge often requires a unique subset within ordinary language. All specialists have their own codes, unfamiliar to the uninitiated and it sets them apart from everyone else. A good example is the use of Latin for medical purposes. Ordinary people usually have common expressions for the same thing, but it lacks the precision and universality of the expert's version. This is a natural development and essential for establishing order in spreading and utilizing information within society. Children learn natural languages in school but even in specialized education they will use languages, albeit possibly in the form of formulas and physical or social laws.

Conversation is a representation process, such as someone relating to others what he/she has experienced; it describes real issues but it is not reality itself. Relating the conversation to someone else again later turns it into a description of the reality of the conversation, etc. These days nearly all knowledge and experience is many steps removed from reality, thanks to our media culture. We are able to know very much more than our ancestors, but what we know is progressively further removed from reality. People often assume they know what experts are talking about, but they may not be getting the entire picture, missing safeguards, exceptions, caveats, etc, which reminds us (me included!) that a little knowledge is dangerous!

Classic knowledge and Civilizations
Logic awakened European enlightenment and its moral conscience

There was an immense decline of civilization and culture in Europe, after the fall of the Roman Empire in the fifth century, AD. This continued for more than 300 years and it is clear that at that time more antiquity artifacts and books survived than there were people who knew anything about it or able to read them. Civilization barely existed during this long period of dark chaos and unrestrained fighting, marked by barbarism, cruelty, and a lack of compassion. The new Christian Church regarded protection of its religion its only duty; protecting people from brutality or enslavement was delegated to warlords, who in return propped up the Church. Then suddenly three significant events occurred in the seventh and eighth centuries, initially unremarkable and indistinguishable from other chaos.

The first was the founding of Islam by Mohammed in Mecca, initiating a march of religious conquest in the Middle East, then North Africa, and finally Spain. Arabic culture flourished and contributed greatly to a revival of global collective knowledge. When Arab fanatics crossed the Pyrenees from Spain into Aquitaine its duke requested help from his northerly neighbor Charles Martel, the effective ruler of the Franks. Martel defeated the Arabs at Poitiers in 732, forever halting the spread of Islam into Europe. It established the foundation for Frankish power, crucial to the future of civilization in Europe.

In 771, Charlemagne (Charles the Great) became the sole ruler of the Franks and he subdued all other kingdoms and tribes, thereby bringing the Christian pope under his protection and breaking the tie with the Byzantine empire. Although he was an ultimate warlord, Charlemagne recognized the value of education and issued a far-reaching proclamation that established a program of scholarships for enlightenment and the encouragement of all the arts. His court in Aachen became a center of education, and not just for the aristocracy. He especially criticized "uncouth" letters from monks and forced the clergy to become better educated and establish various centers of Christian study and teaching. He himself could read but never learned to write. In short, his efforts resulted in large-scale copying and distribution of original ancient literature that could still be found in places at that time. To emphasize its significance: only a few of these original documents exist today; almost all others are copies, ultimately attributable to Charlemagne! He rescued a very large number of ancient documents for posterity.

The real emergence of western power, civilization, and science dates from a period called the Renaissance or European Awakening (roughly the 16th century). Although initially centered in Italy, it equally affected many other regions. Science no longer served the Christian Catholic church but became the domain of individuals, seeking enlightenment and satisfaction of their intellectual curiosity, based on doctrines of inference and reason. This influenced all facets of life throughout the second half of the second millennium, especially in the arts and sciences but also in religion (the Protestant Reformation was a direct outcome). Prior to this there had been many periods all over the world where human knowledge surged and then again declined, after destructive wars, natural disasters or epidemics. But enough classical knowledge had now been rescued, sometimes by the slimmest of odds, that post-medieval thinkers were able to initiate a new beginning by questioning, resurrecting, and consolidating all that came before.

It was the birth of modern global science, including mathematics, philosophy, medicine, earth and life sciences, astronomy, etc. There suddenly were many scholars, self-taught experts in the ancient and classical languages who published translations and interpretations still available today in universities and major libraries. Thanks to Charlemagne and the monasteries, they rescued the old civilizations

(Ancient, Greek, Roman, Arab and Chinese) from the 'dark ages' and the significance of this for us today cannot be overstated. Of crucial importance was the development of the printing press by Gutenberg in Germany (~1439), assuring the critical distribution of enlightenment in Europe. Although many of these early scholars were monks they did not necessarily follow directions from their church superiors, communicating instead with equals all over Europe. An important early example of this was Erasmus, born in Holland but even at that time looked upon as a citizen of Europe with connections in all the leading western countries.

The Protestant Reformation
16th century reason questioned ancient dogma and superstitions

As a humanist and theologian, Erasmus played a crucial role during the first half of the 16th century in reforming the Christian religion. He remained a Catholic all his life, but at the time many blamed him not only for helping but even instigating the Protestant Reformation. It was a tribute to Erasmus' skill as a writer of skeptical literature, by never crossing some obscure but critical line, that he died a natural death at the age of 67 in 1536 (unlike his friend Sir Thomas More, who refused to accept Henry VIII's break with Catholicism). Religious skepticism and criticism was not tolerated lightly by either Protestants or Catholics at the time. When the Protestant reformation eventually got out of hand, resulting in the ransacking of many churches, all of Erasmus' books were banned for a while after his death. Initially a great admirer, Martin Luther became a severe critic and called him a coward and worse, primarily because Erasmus did not join his reform.

However, and to his credit, Erasmus never joined anything and this is probably why he strongly influence all sides. His stated objective in life was to be an independent scholar. He excelled in the knowledge of classical languages and his collation and translation of (Greek) documents into Latin were circulated throughout Europe, as were many other books. He put a humanist bias on it all and greatly influenced the new Protestant movement and the Catholic response to it. In the early 16th century he was responsible for ~15% of all book sales and he was at the center of the literary world for a long time. In strict terms he was not a scholar but the first large-scale writer and editor of social and religious subjects suitable for a large audience. His collaboration with the Basel printer and publisher Johann Froben made both of them famous throughout Europe.

After his parents died from the plague (he was about 15) Erasmus entered the monastic school system, and thoroughly hated its strictness. It colored his attitudes towards the church for the rest of his life. It is an interesting observation that most people who have influenced civilization, like Erasmus, were motivated to some extent by their aversion to authority. Poverty forced him to enter a monastery and he was eventually ordained as a Catholic priest. His intellect was recognized early and a local bishop sponsored him to study at the University of Paris, a center of Renaissance humanism. He never returned to his religious obligations.

'The Praise of Folly' (dedicated to Thomas More) is still a literature classic. It is a satirical treatise of what Erasmus regarded as the irrational aspects of the aristocracy, monarchy and the clergy; it was his most popular book then and now. Another major religious question at the time (and today) was 'free will' and he advocated the view that God does not often influence choices made by individuals, but this contradicted reformist doctrine. This argument comes down to the question whether individuals are free to conform to the ethics of Jesus or if only God can save you from sin.

Possibly Erasmus' biggest contribution to civilization was the first careful translation (Greek into Latin) and editing of the New Testament. It caused much controversy and replaced many incorrectly translated sections in use at the time. He resided in many countries, including Germany, Italy, and England where he was a professor of divinity at Cambridge University and made friends with many of King Henry's intelligentsia. Although the history books describe him as a brilliant philosophical scholar, there was something more inexplicable about him that is no longer apparent to people in the twenty-first century, looking back. He had genuine faith and loyalty but was very critical of Catholic priests and others who did not put the Christian message ahead of church influence and riches.

He was his own man, a pragmatist keenly aware that tearing down the old could be fatal to civilization. He never used his preeminence to enrich himself by choosing sides and instead made everyone mad for not doing so. But he succeeded in maintaining a balance that was in the best interest of all at a crucial time, and he contributed to changing what had been a religious power monopoly. Erasmus was compelled to speak out; and he had the intellectual and moral strength to take a leading role, next to Martin Luther, in delivering a message that drastically reshaped civilization in Western Europe. Erasmus never intended to destroy the Catholic Church, only seeking to limit its excesses. He thoroughly disapproved of Protestant extremism and openly said so.

Rationalism, Facts and Myths
Technology accelerated rationalism but is it leading somewhere?

Religious concepts are often simplified with symbolism because without it they are convoluted, subjective, and inappropriate for the uninitiated, inviting unanswerable questions. All religions advocate faith without questioning as the path to enlightenment, but this has caused many problems in the history of collective knowledge. The pursuit of natural and physical science was often discouraged, banned, or punished, sometimes with death. Religious leaders feared a downward slide in people's faith, diminishing God's (and their own) unlimited authority. Modern religions today face a major task to weed out and explain concepts that were always symbolic. It now stands in the way of modern rationalism, threatening to diminish traditional religion. It is not that atheism is winning, because people still have a yearning for religious belief, but they become alienated by pointless formal rigidity. An admission that core beliefs are primary and everything else is human embellishment would change that.

Affecting us all, a most interesting and important question regarding order is the obvious surge in global social homogeneity. Five hundred years ago each culture was an island, penetrated only by destructive conquest. Today, instant communications indoctrinate everyone with the same or at least similar values and knowledge. What has this done to the local worldview, or outlook on life and the world? We now benefit from at least 500 years of rapid scientific, technological, cultural and social evolvement. It is taken for granted that human rights, political and religious freedom, and many other social benefits, are a birthright to be demanded if not comparable to other places.

In earlier times only religion attempted to instill social order in a society dominated by cliques of selfish bullies. To stop this discourse from degenerating into balderdash, we may conclude that public information today (including scientific) is more factual than mythical, very much the opposite of how it used to be. Such facts are now much closer to reality, where myths vary between families, villages, and cultures. This is the reason for the unstoppable explosion of order, knowledge, and social homogeneity. The questions are: has religion and education kept pace, where is it taking us, and do we want to go there?

CHAPTER III - <u>ENERGY AND SUBSTANCE</u>

HISTORY OF PHYSICAL SCIENCE
**In the absence of any other proof, the thumb alone
would convince me of God's existence.
(Sir Isaac Newton)**

<u>Atomic Interactions</u>
Society is now fully dependent on sub-atomic
technology

This chapter (Energy and Substance) is not a summary, critique, or lecture, on the status of physical science. Instead, and in line with the theme of this book, it asks questions many of which have no definitive answer, suggesting incomplete knowledge. Before you start, please don't feel obliged to read it if you are not curious about physical science. However, it underlies all physical uncertainties and therefore I suggest scanning at least the conclusions. Even when wrong, there is still the mystery! It is astounding that scientists have discovered in only one century, and in great detail, that our Universe consists of electrons, nuclei and quarks, photons, etc. They have formulated unique laws of nature that combine particles into ever more complex structures, ending up with humans, black holes, and everything else in between. Therefore, it is no surprise there has been talk about 'the end of science' for more than a hundred years now. However, this ignores the fact that basic natural phenomena such as gravity, mass, inertia, force, electric charge, and especially photon entanglement, are not understood very well or not at all. Scientists can mathematically describe many or nearly all features and relationships, but a fundamental understanding is missing. Partly, this stems from not knowing what elementary particles really are.

However, physicists know a great deal about particle-structures and their interactions, including nuclear structures such as protons and neutrons, and our fabulous 21st century technology is based on it. Chemistry, pharmacology, electronics, optics, biology, nuclear and material sciences all depend on the predictability of atomic and sub-atomic interactions. Therefore, physicists may be excused for not knowing (or caring) what elementary particles really are, there is little money in it. However, the theme of this book is not about what knowledge they have, it is about what knowledge they don't have. Since their test-equipment also consists of particles, it is unlikely that experimentation alone will ever reveal reality below that level, and progress depends on speculation and deduction. Fortunately, there are many clues that must all logically agree in any new theory, and progress is therefore still possible. It may not be possible to fully understand such reality but useful analogies are necessary to proceed. Such speculation and deduction will come from 'thought-experiments', carried out by individuals and verified indirectly in test-laboratories (using equipment often worth billions).

Analogies and Force
Perceiving the reality of force at the lowest level
is impossible

Scientists and laypersons use intuitive analogies for everything they don't understand, sometimes inventing suitable ones to mimic unimaginable realities. Force is a good example, and even little kids know what that is - you push, and it moves! Experts know a lot more because they can calculate the magnitude of any force, based on compressed or stretched molecules or virtual exchange particles and the conservation of energy. However, ask a physicist to explain the real nature of (intermolecular) force and he or she will mention photons, electrons, and electric charges. Ask about electric charge and it is explained in terms of charge-balance, electrostatic fields and interacting elementary particles; a circular argument.

Truly understanding the nature of force, inertia, mass, gravity, and charge requires new conceptual approaches to Vacuum-physics and new analogies to simulate its characteristics. It is probably impossible for us to realistically visualize the interaction between particles in the Vacuum, the mysterious intermediary of energy that all matter stems from. The likely reason is that it does not take place in our local frame of reference. Most scientists are reluctant to address this; they regard it unnecessary and they are right, except that it is like living in a wooden house without knowing that the wood came from the trees in the surrounding forest.

Missing Foundations
Science can predict nearly everything but a foundation is still missing

Introducing the 'Vacuum' seems like we restore something that died towards the end of the 19th century. Einstein's relativity theories dispensed with an 'Ether' completely, as a medium for transporting light (electro-magnetic energy). However, his contemporary Max Born (in 'Einstein's Theory of Relativity') recalls: - *"Einstein in his later years proposed calling empty space equipped with gravitational and electro-magnetic fields the 'ether', whereby, however, this word is not to denote a substance with its traditional attributes. Thus, in the 'ether,' there are to be no determinable points, and it is meaningless to speak of motion relative to the 'ether.' Such a use of the word 'ether' is of course admissible, and when sanctioned by usage in this way, probably quite convenient".* - The Vacuum is not a substance, since it has no structure and no frame of reference that can be applied to it, but it is not empty as is generally perceived. Matter extends into the space that surrounds it by shaping the energy density of the Vacuum, which is the principle for gravitation in 'General Relativity".

The so-called 'Standard' model of mysterious particles, charges, fields and forces lacks a suitable foundation relating it to the Vacuum. The English physicist Paul Dirac mathematically determined that there must be two types of energy: matter and anti-matter. A few years later the American C.D. Anderson experimentally discovered the electron's anti-particle, calling it a positron. Dirac suggested that this anti-particle is more or less a hole in Vacuum-density. There may be other Universes somewhere made of anti-matter; which is interesting but not helpful, since we clearly live in a Universe made of matter. Physicists see good agreement between their experimental and theoretical results and this is a tribute to their mathematical skills and ingenuity but, unless new experiments provide more information, they may not be able to go much further. A promising new approach was attempted with 'String Theory', but even that has hit a brick wall lately. I personally find String Theory mathematically abstruse with its 11 spatial dimensions (sorry! - I meant to say I don't understand it). They probably should be looking for the missing foundation by using analogies that simulate the Vacuum's relational properties, instead of mathematical monstrosities.

Ancient Science and Magic
Science has yielded the gradual exposure of ancient magic

Science has a long history dating back to early human civilization. We can only investigate back in history to the time that writing was invented, but every new civilization based their science on what came before. Scholars in medieval Europe studied Islamic, Roman and Greek documents, and they in turn examined the records from ancient Mesopotamia and Egypt. Early knowledge from China and India also became available eventually. The scientific method (the logical elimination of predisposed biases or inadequate evidence) gradually took hold during the last few centuries, although it was originally pioneered by the Greek mathematicians Euclid and Archimedes, among others.

Science before the 17th century was based on observation and speculation, or religious dogma and magic when logical explanations seemed impossible. Very little of such magic remains, although there are still plenty of inexplicable phenomena. Scientists are not worried as long as they can accurately predict things using empirical mathematical theories. This book takes the opposite approach; it ignores theoretical details and concentrates on missing fundamental understanding, even if it does seem like looking for magic sometimes.

Newton, Genius & Heretic
Newton's discoveries exposed the inherent order in Nature

A good introduction to physical sciences is a brief review of the life and accomplishments of the individual generally looked upon as the first modern scientist. Sir Isaac Newton (1642-1727) was the foremost researcher of physical science of his age, and possibly of any age. For 300 years he has been seen as a scientific revolutionary, matching experimental and mathematical proficiency with innovation. He is famous for discovering several fundamental relations (math. equations) that link matter, force, motion, and gravity. His discoveries in mathematical calculus (he called it 'fluxions') and optics are a close second. He changed science from a collection of unrelated facts and guesswork to a discipline. We can also make a comparison to Einstein, who is equally famous for his Special and General Relativity (the latter covers gravity). Newton did not always publicize his discoveries, but eventually they did appear in his books. First there was 'Opticks', then the momentous 'Principia' and also his many letters to the Royal Society of London. He became a Fellow (and later President) of this Society in 1671 and then Master of the Mint in 1699, a prestigious post. Queen Anne knighted him in 1705, primarily for his currency innovations. Presumably his scientific achievements were less important to her.

Newton was born prematurely and did not have a happy childhood; he hated his stepfather and had a grudge against his mother for marrying him. He had a dwindling love affair at 19, possibly the end of romance for the rest of his life. His outstanding scholastic performance gained him admittance (probably free) to Trinity College in Cambridge where he received a degree in 1665. Avoiding the Plague, he spent the next two years at his mother's home developing theories on optics, gravitation and calculus. 'Principia' was published in 1687, defining his three universal laws of motion and the law of gravitation. Of great importance to Newton and his achievements was the access he had in Cambridge to the printed translations and interpretations of ancient and classical science-literature, translated by medieval scholars. You probably assume that Newton was a modern man, far ahead in his attitudes, etc, but let's take a look at some lesser-known facts. He indeed displayed amazing logic in a superstitious age. However, encountering issues he could not rationalize he tended to fall back on a mixture of medieval occult (the opposite of science), religious anxiety, alchemy, and pure magic. A large portion of his unpublished nonscientific notes were purchased by John Maynard Keynes, the economist, and we can learn a great deal about 'Newton, The Man' from a Keynes' lecture with that title, given in 1946.

Newton viewed God as a designer, rather than a participant in life or manager of the Universe. His reputation and writings were influential in convincing intellectuals all over Europe that humans determine their own destiny by decisions they make and that they are collectively responsible for most evil. This was a concept with far reaching consequences in civilization. Newton was a very difficult individual and did not have many friends, but the few he had were prominent in the Royal Society and they were convinced of his extraordinary talents. They used him as their antidote to the chaotic and confused philosophical views of the time, originating from religious conflict and remnants of medieval superstitions. He was their champion, and Newton managed to fulfill that role because his intellect made him see through much of the prevailing superficial irrationality. However, behind that facade he was mystified by the many obvious riddles in biology and nature.

Although never publicized during his lifetime, he expended enormous effort writing about dubious and risky non-scientific subjects and he ventured opinions deemed heretic at the time. Keynes describes a few attempts in later centuries to analyze and publish some of it, but religious authority always judged it harmful to society. This is not surprising; even today there are many people willing to accept occultism to satisfy superstitious intuitions and Newton still has popular appeal, assumed to have had special insight, like Einstein. We forget sometimes that he was unaware of all that Science has discovered since that time.

Newton held views not dissimilar to that of modern creationists. Had he lived today, his logical intellect would have steered him away from it because of irrefutable evidence provided by Evolution Theory and Paleontology. His religious convictions were entirely effected by indoctrination, his emotional disposition, and a lack of the information we take for granted today. Newton was a phenomenon, a bright light in the dark, a socially awkward and suspicious genius of extreme contradictions and talent, and (quoting him) 'standing on the shoulders of giants'. He logically observed Nature and the Universe but at the same time glancing back to the medieval past for magic.

Science in the 17-19th Centuries
Search and expectation of perfect order in Nature and Society

The period after Newton, until the beginning of the 20th century, was a golden age of new and rational discoveries. In addition to Newton it was based on the work of many, in particular the 17th century scientists and mathematicians Descartes, Huygens, Pascal and Leibniz. It gradually became obvious that there is predictable order in nature, and it eventually led to an arrogant overconfidence; but that will be covered later. During the 18th and 19th centuries there was a burst of experimentation, resulting in many new scientific theories. Franklin, Dalton, Young, Fourier, Faraday, Laplace, Joule, Thomson, Pasteur, von Helmholtz, Clausius, Boltzmann, Gauss, Weber, Hertz, Darwin, Mendeleev, Lorentz, and Mendel are all names of scientists associated with the foundation of Physics, Chemistry, Biology, Mathematics, etc. The brilliant British mathematician James Clerk Maxwell who in 1873 combined light, electricity, and magnetism into the theory of electro-magnetism culminates this period. After that it was thought, at least for a while, that all remaining uncertainties were minor details. It was a science of tiny billiard balls with fields of energy gradually running down into chaotic heat.

Cracks of dissent already appeared when the American Michelson and Morley experiments failed to detect the existence of an 'Ether'. Further doubts emerged with the discovery of electrons, X-rays, and radioactivity, after exhaustive investigations by many researchers such as Thomson, Rontgen, Becquerel, the Curies, and Rutherford. The lack of scientist-names from outside the Western world is deceiving; many talented people participated but had difficulty to be recognized in their own countries. Scientific applications and technological development could not be stopped and the number of patents worldwide exploded. Steam, diesel and gasoline-engines now powered factories and transportation. Medicine and chemistry, along with everything else, were also reformed in the great Industrial Revolution. Probably the biggest impact on society was the availability and application of electricity, promoted by inventors such as the Americans Edison and Westinghouse, and it changed commerce and the way of life for almost everyone. Communications via telegraph and telephone were seen as the ultimate in human inventiveness, only to be overtaken by radio, using electro-magnetic waves in a mysterious Vacuum. Of course, human nature and prejudices had not changed and such technological advances only increased the desire and ability by some nations to dominate others. The resulting wars, fought with more people and new weapons, were bloodier and more heartbreaking than ever before. It disgusted everyone, but patriotic pride in the millions that perished overpowered it. Who was to blame seemed unimportant; the German Kaiser received house arrest for sanctioning the first World War. This war, and the next, are good examples of the extreme peril created by uncontrolled human conflict between logic and instinct.

The 19th century had closed with a sense of exhilaration and accomplishment, at least in the scientific community. However, it did not last long and problems should have been anticipated by anyone who paid attention to the growing list of scientific loose ends that had been swept under the rug. Physics, as the 'fundamental science', would suffer the brunt of coming changes. Still, the unification of light (radiation) and electro-magnetism (defined mathematically by Maxwell in 1873) convinced everyone that the foundation of physical science had been discovered. With Dimitri Mendeleev's periodic table of chemical elements, and Thomson's discovery of the electron as the carrier of electrical energy, the remainder seemed only minor details.

ENERGY QUANTA AND RELATIVITY
The human mind has first to construct forms, independently, before we can find them in things.   (Albert Einstein)

A Scientific Revolution
A collapse of certainty and of confidence in
classical physics

During the first quarter of the 20th century the general public became aware that the scientific community was taking some outlandish concepts very seriously. However, it took decades before it sunk in that this was not just the excessive imagination of a few scientists. These notions were incomprehensible, even for many professionals, but it fueled the general public's significant interest in the paranormal at that time. It started a semi-scientific cult, continuing to this day as evidenced by a flood of books and magazine articles on such subjects. First there was Max Planck, with his theory of only discrete units of energy (quanta) radiating from hot objects, and then Einstein with his perplexing theories of Special and (later) General Relativity. Also Niels Bohr, and the complications he encountered with electrons instantly jumping from one orbital energy level to another in his otherwise successful model of the atom.

There was slow acceptance that atomic particles were not like little billiard balls, but fuzzy patterns that changed energy level only in specific amounts (quanta) and were subject to uncertainty and statistical probability. There were many other scientific heroes who managed to get their name in the history books. All this shattered the comfortable attitude that Classical Physics (prior to 1900) and the laws of nature could predict everything. 'Relativity Theory' and 'Quantum Mechanics' caused immense upheaval over the next fifty years, leaving classical physics as a limiting case only. High School physics is still applicable for ordinary (macro) objects but, under extreme conditions and certainly when dealing with elementary particles and radiation, the laws of nature are more complex.

Many things were discovered, none more fundamental than Einstein's $E = m*c^2$ equation. It means that all matter has energy locked up in its structure, potential and kinetic, so much that 1 kg. of matter is equivalent to 21,000,000,000 kg. of TNT, sufficient to put a mountain in orbit if it could be released! Alternately, it means that any concentration of energy has mass and/or momentum and is subject to inertial forces, gravitation, etc, and this is a very important aspect of both relativity and quantum mechanics. Physicists today have to make sense out of about 6 basically stable and at least 16 unstable (short lived) particles and their anti-particles, a bit like watching a soccer game with 22 players without ever seeing the ball. Clearly, they have not yet discovered the root-source of our physical world and they may never, except by deduction and analogy! Whatever arranges energy into these configurations may be theorized, predicted or calculated, but it will never be sensed. Our test-equipment is incapable of detecting energy-fluctuations at such a low level unless it is made of building blocks much smaller than atoms.

Einstein and the New Physics
Einstein was ultimately ignored because he went
too far, too fast

Albert Einstein (1879-1955) was a 20th century cultural idol, not of his own making but by circumstance. People perceived him as a scientific prophet, primarily because he and his supporters could not explain the Special and General Relativity theories in a simple enough formation, making it appear like magic. He was born in Germany but became a citizen of Switzerland in 1901 to avoid German military service, switching back again to Germany in 1914. He became a citizen of the USA in 1940 after leaving Germany in 1933 for political reasons, like many other Jewish people. His childhood was ordinary, but the rigid educational system of the day did not suit his temperament. He had an aversion to authority, needing to follow his own path. This caused him much difficulty, although he was clearly exceptionally intelligent. Only in University, training as a teacher of physics and mathematics, did he submit to the rigors of serious study. He eventually found a job in the Swiss patent office in Bern after receiving his degree in 1902.

Over the next three years he received a doctoral degree from the University of Zurich in physics and prepared three articles for publication on 'Special Relativity', 'Brownian (molecular) Movement' and the 'Photoelectric Effect'. Although not immediately recognized as such, all three will forever be prime examples of the human ability to utilize insight, intuition, and logic to hypothesize conclusions based on observations. Einstein was reasonably skilled in mathematics, but what he called 'thought experiments' were his real strength. Almost every revolutionary concept in physics during the 20th century is based on Relativity or Quantum Theory, and Einstein originated the first and was a major participant in the second. He completed his Special and General Relativity theories single-handedly but received a Nobel Prize (1921) for his interpretation of the photoelectric effect, which provided the all-important firm basis for development of Quantum Mechanics.

His new concepts of physics became well known by around 1908, and his reputation increased. His marriage (he had two sons) had failed due to incompatibility. He became a professor at the University of Berlin and started work on extending his Special Relativity Theory to include accelerating objects. This led in 1916 to the revolutionary theory of 'General Relativity', treating gravity as a distortion of space-time by matter. His prediction of deviations from Newton's gravity-theory were confirmed after the First World War and it made him an instant celebrity until his death in 1955, and probably forever in human history.

He argued with the physics community about quantum theory, its uncertainties and probabilities, because some features clashed with Special Relativity (e.g. instant action at a distance). He had a very public debate about scientific determinism or causality with Niels Bohr, the Danish physicist leading the statistical probability camp, and everyone at the time assumed Bohr won the argument with what is called his 'Copenhagen' interpretation. However, Einstein continued to insist that quantum theory was incomplete and many people now agree that he was right. He spent many years attempting to unify all fields and forces of nature but this was bad timing because not enough was then known about the nucleus structure. Although there are underlying relationships, and there has been progress, attempts to unify all of physics have not been successful so far.

Einstein had a significant impact on the entire world, and not just scientifically. His letter to U.S. President Roosevelt in 1939, warning of the potential development of a nuclear bomb in Germany, launched atomic fission research in the USA. Einstein agonized over it, after Hiroshima and Nagasaki, feeling doubly responsible because nuclear weapons are based on principles he unveiled forty years earlier. It repeated his earlier experience in Berlin during the first World War when he was at odds with 99% of his acquaintances over the morality of this war. Then, when asked to contribute with an article for a national organization about the 'rightness of the war', he responded with something like: - Honor your master Jesus Christ, not only in words but by your deeds.

Motivated by logic and principles he had a rather lonely life towards the end, surrounded by scientific, political, and social controversy. The FBI collected a large file on him during the absurd McCarthy era because of critical socialistic comments he uttered occasionally. It represented a devastating and frightening verdict against the evil sway of self-righteous and obsessed politicians.

Einstein was an active Zionist for many years, cultivated by Jewish organizations which he supported to the best of his ability. However, his religious views are summarized by the following (1950) quote: - *"My position concerning God is that of an agnostic. I am convinced that a vivid consciousness of the primary importance of moral principles for the betterment and ennoblement of life does not need the idea of a law-giver, especially a law-giver who works on the basis of reward and punishment."* - He considered organized religion man-made and hated the need for military force; but he loved mystery. He was never the ideal husband and father, always putting his obsessive interests far ahead of his family, but still claiming that one exists for others. He was a loner, especially to his family, but uniquely capable of intellectually interpreting everything, including morality and compassion.

Non-locality and Reference Frames
Relativity theory signaled the first violation of
self-evident logic

To the uninitiated, physics is schizophrenic. Nothing goes faster than light but, in an entangled pair of photons (created simultaneously in a single event), one will change its spin-direction instantly over an enormous distance whenever its partner is changed. Physicists take this in their stride; it depends on how you look at it they say, and besides, non-locality (instant connection over cosmic distances) is a basic premise of quantum mechanics. Yes, but that is not an answer for anyone, except for physicists with religion-like faith! The Big Bang theory doesn't work unless there was an initial period of inflation at a speed billions of times faster than the speed of light. A key piece of information here is that at this stage no elementary particles had yet formed. However, inflation theory worked and was embraced by almost everyone, eventually!

A major clue is Einstein's ('Special Relativity') postulate that the speed of light in a Vacuum is a true constant and even an astronaut traveling at half lightspeed (if that were feasible!) measures the same incoming speed of light from all directions. The major consequence of this theory is that the laws of physics are the same at any speed and in any reference (coordinate) frame. Experiments confirm that an astronaut's clock slows down because all parts (including a balance wheel, if it had one) gain mass with acceleration, relative to observers on Earth. However, our infallible logic dictates that light coming from stars ahead and behind him/her can't possibly have the same speed, but it does (according to both Special Relativity and experimental evidence)! Something is therefore drastically wrong. This 'something' is our misconception of the nature of energy at its most basic level, and the lack of a frame of reference.

Special Relativity features time/length dilation and mass/energy equivalency but more than laypersons have misconceptions, because quantum and relativity theories have never been combined into a single theory. The late John Bell, a quantum theoretician famous for his 'Bell Theorem' and a supporter of David Bohm, made the following comment: *"It may be that a real synthesis of quantum and relativity theories requires not just technical developments but radical new conceptual renewal"*.

Our astronauts know how fast and in what direction they are travelling because it shows in the average (Doppler) frequency shift of starlight coming from in front and behind. Then how is it possible that the speed of this incoming light is the same? The physicist David Bohm gives us a key to this puzzle by proposing that particles form and dissolve at a pulse-rate very much greater than light-wave frequencies. Each particle is at rest at the center of its own (Vacuum) reference-frame, radiating and reforming rapidly and tracing a discontinuous trajectory in time relative to others. Assuming relativity doesn't make much sense to you, try the following analogy:

  - *A huge solid rubber ball with vibration-sensors at its center moves at a high constant speed relative to stationary machine-guns, repetitively firing at it from all sides. The direction of motion and velocity relative to each machine-gun is sensed because the interval between bullets hitting the rubber from any direction is inversely proportional to velocity. However, the velocity of vibration waves in the solid rubber ball is the same from all directions -*

With that we can understand why astronauts sense a constant speed of all starlight, because, like in the rubber ball, waves propagate in Vacuum at the same speed from all directions relative to each particle (or system of bound particles). The Vacuum is continuous and lacks a ref. frame; it always appears to be at rest relative to waves emanating from any (pulsating) particle. The difference is that we understand the medium 'rubber' very well and the medium 'Vacuum' not at all! Don't take the above analogy too literal, it is a representation of the constant light-velocity phenomena. To match Einstein's equations you would have to assume that the rubber ball, with Vacuum properties, extends right up to the machine guns. This is the closest you may come to a (macro) physical explanation for Special Relativity's bizarre Vacuum-attributes!

Continuous Vacuum and reality
The Universe consists (only) of vacuum-density fluctuations

The analogy in the previous section suggests that the energetic Vacuum is real, and not just a fallacy to explain the unknowable, or a convenient theoretical crutch as Max Born suggested. The key is that whatever external stimuli influence the astronauts they only sense it in their own frame of reference, or their own Vacuum-coordinates. This is fundamental to everything that is weird about Einstein's theories. It remained a mystery and he became a cult-hero to many people, believed to have some sort of esoteric vision. Although he tried, he was unable to provide the general public with a simple illuminating analogy they could understand. A section called 'A Vacuum Analogy' is proposed later in this chapter and it provides a unique opportunity to visualize particles, fields, and interactions, but don't think of it as reality, because no one knows what Energy is!

It assumes that what we call particles are quantised (discrete) states of the Vacuum and what appears to be empty is actually the continuous energy (non-discrete) state of these same cyclic particles. Two bodies moving at different speeds receive photons from each other at absolute lightspeed. This seems physically impossible and the only conclusion can be that the clocks of the two objects are not synchronized. This is nothing new; it is a fact, and now you know why: the Vacuum is continuous and has no relative velocity to either object. Each particle or object (or celestial body) creates its own energetic Vacuum! Problems only arise in our mind; it is our inability to visualize something without a frame of reference that is responsible. From the analogy we draw a fundamental conclusion: 'Planck' quanta may be the smallest energy quanta partaking in energy exchanges, but the Vacuum itself is continuous. If not, it would have a ref. frame and we know it does not. Physicists talk about a 'Quantum-Vacuum' but that makes little sense; they probably mean that the Vacuum can form energy quanta. With sufficient local energy-density the Vacuum has the potential to create quanta of energy, virtual, or real when the conditions are right.

The crucial revelation by Dr. Einstein was that any object approaching lightspeed increases in (relative) mass. This is humdrum for physicists now because $E = M*c^2$ applies always; for instance, proton and neutron masses are much larger than an electron's mass because quarks and gluons (the constituents of an atom's nucleus) zip around at close to light-speed. Almost all of their mass is kinetic energy (momentum), much more than their 'rest mass'. Mass of an object increases during acceleration relative to a stationary observer! Quarks gains mass temporarily when (virtual) gluons accelerate them, absorbing and conserving gluon energy; each nucleus acts basically as an intricate and fast push/pull process. The same happens when electrons are accelerated. Objects absorbing a photon increase their mass very slightly, conserving the photon's energy. When a photon is absorbed by an electron it makes the latter more energetic and increases its mass relative to external observers, but loses mass when it emits a photon. The entire system of matter and radiation (the Universe) is a gigantic and chaotic movement of Vacuum density fluctuations, interacting by exchanging and conserving energy.

Special Relativity and Time
Time-travel into the past is irrational and impossible

Television programs about Special Relativity like to show the weird effect of objects getting shorter when they approach lightspeed. This is of course not true when observed in the object's frame of reference. It is primarily a virtual effect since measurements use light beams and time and we now know both are not absolute. It is still interesting to understand why this happens since it is fundamental to the concept of relativity. Contraction in the direction of movement was predicted by Fitzgerald and Lorentz when Michelson and Morley discovered in 1887 that lightspeed had a constant value (impossible in classical physics). Eventually Einstein told the entire story to a skeptical world in his 1905 Special Relativity Theory.

Length contraction and time dilation are not mysterious, assuming there is a constant lightspeed in Vacuum. But such constant lightspeed is counterintuitive and a consequence of $E= M*c^2$. It will be addressed later in a section entitled 'A Vacuum Analogy'. Accelerating any object increases mass cumulatively until it nears lightspeed and further acceleration then becomes impossible, requiring infinite energy. Length contraction shows itself when an astronaut, moving at high speed, reflects a light-beam off a mirror (at 90 degrees to the direction of travel) back to its source. With constant lightspeed its path appears longer (a 'V' shape) to a stationary observer, due to the velocity, and therefore takes more time (the astronaut's clock is slower!). Also, with a slower clock, a meter-bar measured in the travel-direction with a light-beam will appear shorter to the stationary observer.

Changes in clocks and mass are not virtual but have a real physical effect. For instance, protons traveling at relativistic speeds can harm us as cosmic rays during space-travel due to their very large mass (energy). When a spacecraft from Earth decelerates (accelerates in the opposite direction), its clock-rate (relative to Earth) will go back to normal when reaching its original (Earth) speed. However, continuing to accelerate in that direction does not make the clock go backwards. You are supplying energy (relative to Earth) and the clock slows down again; the 'arrow of time' has only one direction, into the future – slow or fast, but never back. Time travel into anyone's past is impossible, but you could meet someone else in his/her future, but not your own (similar to being frozen for a while, and what fun is that?).

Biological time also stretches because an accelerated astronaut's molecules, etc are more energetic than before! It makes a bit of sense because, as mentioned, you can easily visualize someone flash-frozen and awoken 10 years later, at least in concept. Time dilation is similar, except body processes do not stop, they just slow down (relative to someone on Earth). It is therefore irrational to suggest that you could do the opposite and wake up 10 years earlier. Along with particles, time was created during the Big Bang. Time does not exist in a continuous Vacuum; it only exists for (discrete) matter and has nothing to relate to without particles. Einstein developed his entire theory by assuming that the speed of light does not vary in any ref. frame; it was an absolutely marvelous, correct, and stunning concept at the time. Photons travel at the speed of light relative to both the emitter and receiver even when these are moving towards or away from each other. This is only possible if their internal clock-rates differ, allowing photons to travel at lightspeed relative to both. OK, let me see if this diatribe can be concluded without mentioning the words 'relative' or 'frame of reference' again, although they are the key to Einstein's theory!

Attempts to apply reference frames to the Vacuum have changed over the last few centuries, switching from the entire Universe to macroscopic bodies, and now to elementary particles. The Dutch physicist Lorentz (one of Einstein's foregoers on relativity) always believed the ether had a single ref. frame, but that concept disappeared with Special Relativity. The true nature of time and space is beautifully revealed by cosmic rays, coming from outer space. Originating from exploding stars, these very high-speed particles are mostly protons breaking up in the Earth upper atmosphere into pions, which then decay quickly into muons and neutrinos. Muons are similar to electrons but 205 times more massive. None of these particles are stable and muons decay after traveling about 2 millionths of a second in experiments on Earth but in this case, due to their extreme velocity, time dilation allows them to exist up to 16 times longer. This is a perfect example of a particle following the beat of its own drummer since the average decay time for an unstable particle is always the same in its own ref. frame. However, time dilation stretches it for observers on Earth.

Planck and Quantum Theory
Proposing that energy comes in tiny packets
caused a revolution

Classical physics treated matter-particles as electrically charged points and electro-magnetic radiation as waves in electric fields, and everything was continuous and unbroken. 'Quantum Mechanics' in the first half of the 20th century changed this to discontinuous and discrete. The history of this revolution spawned multi-thousands of textbooks and biographies. The revolutionaries are legends from a time when mathematicians did not quite have the upper hand yet, although they certainly do now!

It is ironic and fascinating that one individual responsible for proving that 'discreteness' was real, against enormous opposition, was Albert Einstein. And yet he went into history as resisting the uncertainties and probabilities that became such a major feature of Quantum Physics. In my opinion Einstein was basically correct, but he should have been less emphatic after realizing that the information he needed was not yet available. It was in character that the world's foremost scientist (named man of the century in 2000) risked his reputation by opposing a large majority, out of conviction but without proof! An internet quote from 'RELATIVITY' by Albert Einstein, 1952 (Random House) states:

"In conformity with the present form of the quantum theory, (the present-day generation of physicists) believes that the state of a system cannot be specified directly, but only in an indirect way by a statement of the statistics of the results of measurements attainable on the system. The conviction prevails that the experimentally assured duality of nature (corpuscular and wave structure) can be realized only by such a weakening of the concept of reality. I think that such a far reaching theoretical renunciation is not for the present justified by our actual knowledge, and that one should not persist from pursuing to the end the path of the relativistic field theory."

First of all, to make sense out of Quantum Mechanics it is mandatory (and interesting) to review the more significant historical events leading up to it. We can use hindsight and speculation to obtain a clearer picture.

It all started with the German scientist Max Planck who (in 1900) published a paper proposing that radiant energy (light) is not continuously and infinitely divisible, as had been assumed. He was motivated by an absurd theoretical prediction of infinite energy at the high frequency end of the electro-magnetic spectrum. Planck showed that the only reasonable explanation was that radiant energy comes in small packets, using a statistical approach developed earlier by Ludwig Boltzmann. The idea was highly unpopular because it contradicted the very core of classical physics, and even Planck did not accept it wholeheartedly. However, it motivated Albert Einstein in 1905 to publish a scientific paper on the photoelectric effect (discovered 20 years earlier) with proof that light consists of particles, later called photons. This effect occurs when light falls on the surface of some metals, causing electrons to be ejected. They are ejected only when the electro-magnetic radiation exceeds a certain frequency, after which the rate of ejection is proportional to intensity. Very intense light at a slightly lower frequency will not eject any. This only made sense if light consists of energy packets and not waves, as was accepted by every scientist in the world at the time. Such an energy packet or 'photon' has energy proportional to the frequency of the electro-magnetic radiation (light). The proportionality constant is called 'Planck's Constant' (h), a fixed number.

<u>Waves and Particle Duality</u>
Only Quantum Mechanics predicts a stable
atomic structure

All scientific knowledge of electro-magnetism in 1905 was based on wave-theory. In fact, Young's two-slit experiment in 1803 had proven conclusively that light is a wave phenomenon, reverting back to Huygens and tossing out Newton's particle theory. This is the background prelude to probably the biggest scientific battle in history which ended only six years after Einstein received a Noble prize for his new interpretation (in 1921, and 3 years after Planck received his). From that time on, all difficulties were swept under the rug with the assumption that light was both waves and particles (wave-particle duality), and you were not allowed to insist on reality and ask too strongly how that could be. The official position still is that the exact location of a photon cannot be determined; it has a statistical distribution and only the probability of finding it in a certain location is knowable. That does not quite explain it to anyone looking for reality and many people suggested different rationales but were ignored. You will later encounter yet another one because questioning a mystery is natural, except for scientific lawmakers.

Another nail in the wave-theory coffin was a proposal in 1913 by the Danish physicist Niels Bohr to solve a mystery associated with a new model of the atom. The New Zealander Ernest Rutherford proposed it in 1911, based on scattering experiments with helium nucleii, proving that atoms have a tiny nucleus with a positive charge and negative electrons in its orbit. The mystery was that classical physics predicts that such electrons should lose energy in orbit, but this was obviously wrong because atoms are stable. Bohr proposed to take Planck's quanta into account and therefore electrons do not lose or absorb energy continuously but only in small (quantum) steps.

Any electron absorbing a quantum of energy (a photon) wants to 'jump' to a higher orbit, only to be pushed back immediately by photons from electrons already there and forcing it to release yet another photon as surplus energy. The emitted frequency (energy) of this photon depends on the electron's orbit-level. It neatly explained the unique spectral frequencies for chemical elements in thermal radiation coming from hot substances. Heat is simply the kinetic energy (momentum) of atoms vibrating within their molecular constraints, transferred eventually by photons to the environment, or radiated into the Vacuum. Bohr was awarded a Noble prize in 1922 for his quantum concept of the atom.

Probability and Uncertainty
Discrete and non-discrete energy states are the cause of uncertainty

The 1920's was a hectic period when quantum theory was further developed by many participants including Niels Bohr, Werner Heisenberg (Matrix Mechanics and the Uncertainty Principle), Erwin Schrödinger and Louis De Broglie (Wave Mechanics), Max Born (Probability Distributions), and Paul Dirac (anti-matter). Finally, after 1927, all these separate proposals were combined and called Quantum Mechanics. Einstein never accepted the uncertainty and probability aspects of it and concluded (probably correctly) that Quantum Mechanics was incomplete. The new knowledge was used extensively in nuclear research and led to the discovery of protons and neutrons. This period culminated with splitting of the uranium nucleus (fission - the release of surplus binding energy), and the beginning of the 'Nuclear Age'.

Quantum Mechanics successfully predicts particle interactions based on special (Copenhagen) probability rules that deviate from classical physics and it introduced wave-particle duality, the uncertainty principle, the Pauli exclusion principle, and quantum entanglement. The problem with Q.M. is that its originators postulated with religion-like conviction that none of it can be envisioned, making it impossible to grasp for non-mathematical speculators (affectionately called 'crackpots' by the physics establishment). It is better to disregard the lack of reality agonized over in quantum physics, accepting that there is a connection between all energy quanta and therefore they are cyclically extended and distributed in space. If you try to measure how much energy a quanta has, you cannot at the same time determine the location of its center. We encounter the same situation in our macro reality, without ever worrying about it. A speeding driver may be shown a radar picture of his car with the speed printed on it. You could argue (unsuccessfully) that this evidence is inconclusive because it is not possible to determine the speed of anything at one specific moment in time. It requires two time-shifted pictures or a time-interval to measure a 'doppler' frequency-shift and the second picture might show another car, or two cars.

Many people are very uncertain about Q.M. because one of its properties is called the 'Uncertainty Principle'. This is unfortunate because it only means that a collision between an electron and an energetic photon can tell you fairly accurately where the electron is, but only roughly how fast it is going. A low energy photon can provide the electron's speed more accurately, but not its location. You can't have it both ways! Postulating that electro-magnetism and gravity arise from independent origins makes it easier to accept the premise of basic uncertainty in Nature, at any level. We never know with any certainty what an ant will do next, since we don't control it and the ant has no control over us. In contrast, if all things were directed only by electro-magnetism then the actions of the ant should in principle be predictable!

PHYSICS AND COSMOLOGY TODAY
Our view of the particle world is like watching 22 soccer players without ever seeing the ball.

<u>Gravity and the Universe</u>
The force of gravity in General Relativity is a
geometric distortion

Cosmology is the study of the origin and structure of the Universe, and Gravity is what shapes its large-scale composition. Cosmology is now in an upheaval because of two relatively recent discoveries, casting doubts on the foregoing. Firstly, the Universe appears to be undergoing an accelerated expansion, and the popular culprit is 'dark energy', causing mysterious negative gravity beyond galactic distances. The second problem is the confirmed flat rotational curve (equal velocity) of outlying stars in galaxies, although Newton's law predicts that galactic rotational speed should reduce at a greater distance from the center. This discovery is now confidently pinned on the existence of invisible 'dark matter', a substance of unknown composition. Interestingly, dark energy and dark matter have opposite effects, with a negative and positive modification to cosmic gravity, respectively. Science has not yet produced any convincing arguments of what gravity really is (or what causes 'curvature of space-time') and the above two anomalies may be caused by hidden details of a process of which gravity is only one manifestation.

Newton was thoroughly perplexed by gravity three hundred years ago, believing that 'action at a distance' was absurd. Everyone before him had considered gravity, static electricity, and magnetism as pure magic, with major supernatural implications. It did not stop Newton from correctly defining a law for gravity, adequate for all of science until 1915. In his words:

"I have not yet been able to discover the cause of these properties of gravity from phenomena and I feign no hypotheses. It is enough that gravity does really exist and acts according to the laws I have explained, and that it abundantly serves to account for all the motions of celestial bodies. That one body may act upon another at a distance through a vacuum without the mediation of anything else, by and through which their action and force may be conveyed from one another, is to me so great an absurdity that, I believe, no man who has in philosophic matters a competent faculty of thinking could ever fall into it."

Einstein's 'Theory of General Relativity' (1915) changed the concept of gravity, by applying the principles of Special Relativity to accelerating frames of reference. It holds that gravity is a curvature of space-time near any mass (or concentration of energy) - a geometric distortion, which includes bending of light around massive objects and time-variance at relative velocity differences. This worked well and is unchallenged to this day. The force of gravity is obvious on the surface of the Earth when you are accelerated at 9.81 m/sec^2 because in General Relativity's 'curved-space ' only a free-fall towards Earth's center is an 'at rest'-condition.

To question by what means a concentration of energy can cause such distortion in space is rather presumptuous, like challenging how cows make milk. Everyone was happy with statements such as – "The curvature is such that the inertial paths of bodies are no longer straight lines but some form of curved (orbital) path, and this acceleration is what is called gravitation". Einstein himself was a little less satisfied, hinting at a model (full of hidden meaning) where particles are spatially extended and each has a unique frame of reference:

"Space-time is not necessarily something to which one can ascribe a separate existence, independent of the actual objects of physical reality. Physical objects are not in space, but these objects are spatially extended. In this way the concept 'empty space' loses its meaning".

Gravitation and Instant action
Gravity of orbiting bodies acts at lightspeed but in a straight line

Most cosmologists assume that gravity acts at lightspeed because Special Relativity states that nothing exceeds the speed of light. But there is a problem with that, recognized already in 1920 by Sir Arthur Eddington: *"If the Sun attracts Jupiter towards its present position S, and Jupiter attracts the Sun towards its present position J, the two forces are in the same line and balance. But if the Sun attracts Jupiter toward its previous position S', and Jupiter attracts the Sun towards its previous position J', when the force of attraction started out to cross the gulf, then the two forces give a couple. This couple will tend to increase the angular momentum of the system, and, acting cumulatively, will soon cause an appreciable change of period, disagreeing with observations if the speed is at all comparable with that of light"*.

Einstein's General Relativity Theory supposedly solved the problem since motion of an object is affected by the permanent curvature of gravitational fields near any mass, acting instantly. However, the concept of gravitational fields permanently etched in the fabric of space-time is irrational (how can it change instantly when stars, planets, etc. interact with a lightspeed time-delay?).

Using the Earth/Moon example, gravity acts on a straight line between their instant centers, otherwise the moon would have collided with the Earth long ago. But photons are exchanged between them with a 1.3 seconds time delay, causing a small angle between the actions of gravity and light at any instant (we see the moon where it was 1.3 seconds ago). So why question permanent curvature of space and time surrounding concentrations of energy; it works! The answer to this obvious dilemma is that never mind how strange it seems, we have to accept the Special Relativity premise that the Vacuum has no ref. frame of itself and that it effectively rotates along with the Moon-Earth system about their combined center of gravity. Using the rubber-ball analogy, vibrations rotate with the Earth and Moon embedded in a single ref. frame! Therefore, no couple and no problem! Photons on the other hand have their own frame of reference and arrive from the Moon at a slight angle due to their transit time. This suggests electro-magnetism and gravity are not directly related.

<u>Big Bang and Particle creation</u>
Physics' 'Standard Model' of particles is a
triumph of Science

Let's look at how physicists think our world of particles and radiation came into being. If this sounds boring, or too much detail, you should now skip to Chapter IV or V. 'Big Bang' theory proposes that (somehow) extreme density gradients occurred in a highly energetic and unstable Vacuum. It started with the expansion (and squeezing out) of unsymmetrical energy-regions. Next, one such region (containing our future Universe) inflated trillions of times faster than the speed of light (the 'Big Bang'), until the density dropped enough to arrest permanent matter and anti-matter particles (quarks, electrons, and positrons). Most annihilated each other, converting into radiant energy, but what remained (except for present day electrons) formed stable protons and neutrons with their extreme thermal velocity locked up as internal kinetic energy and binding energy (gluons) inside nuclei. Fusing into a three-particle structure reduces their electric charge to a stable level of either -1/3 or +2/3. The answer to why we see only matter and no anti-matter remained a mystery. Anti-matter is just like matter, except with all electrical charges reversed.

Physicists are convinced that we live in a 'matter Universe' and that anti-matter disappeared after the 'Big Bang'. There must have been an equal number of stable elementary particles and their anti-particles initially, but it seems likely that positrons lost out in a symmetry-breaking throw of the dice to electrons. Of course this is controversial; actually it is heresy! Protons and neutrons then eventually captured every positron and most of the electrons, leaving only enough electrons to orbit and electrically neutralize each nucleus; and that created our electrically neutral 'matter' Universe. The capture-process supplied quarks with their large kinetic energy and mass, freezing it at primordial levels. Why do protons have a charge and neutrons do not? The 'Standard Particle Model' assumes that a neutron has two (-1/3) down-quarks and their rest-mass then electrically balances the single (+2/3) up-quark. A proton has two up-quarks and a single down-quark, neutralizing the charge of its external electron.

Physicists have developed a comprehensive model of all (stable and unstable) particles from their experiments and they can predict most interactions, based on experimentally derived constants. This is the 'Standard Model' and it accommodates all known building blocks of the Universe, even under extreme conditions. A huge effort, it is one of the biggest triumphs in Science. There are protons, neutrons, electrons, their anti-matter particles, neutrinos, and a host of unstable particles including force-carrying virtual ones. This Standard Model covers three out of four forces: electro-magnetic, weak, and strong, but not gravity. Each has its own 'carrier' particle such as the massless photon for E.M. force. The strong force binds atomic nuclei and the weak force makes some particle-structures unstable and it is the only force that interacts with neutrinos. This gets very complicated and we better ignore it all for now, especially the details.

Although very successful, the Standard Model has some problems. It predicts the existence of a 'Higgs' particle, undiscovered so far but needed to give rest-mass to all other particles. Furthermore it does not account for 'dark matter' and 'dark energy', believed to make up 22% and 74% resp. of all energy in the Universe. Few scientists worry about dark matter, confident it will eventually be detected. The nature of dark energy is unknown but it is thought to be responsible for the recently discovered accelerating expansion of the Universe, although some people propose that rotation of the entire Universe is the cause of that.

Incomplete Quantum Mechanics
Uncertainty in physics may be caused by the
duality of energy-states

Einstein's biggest problem with Quantum Theory was Heisenberg's 'Uncertainty Principle', endorsed first by Niels Bohr and then by most of the Physics establishment. This principle states that particle behavior or movement is not entirely predictable, even when all initial conditions are known. Particle behavior, and that of any system or group, is only statistically predictable. Einstein disagreed strongly, believing (like many others) in determinism, but he lost the argument at the time. Werner Heisenberg showed that 'Planck's constant' (any quantum of energy divided by its frequency) is always smaller than the uncertainty in particle position times the uncertainty of its momentum (velocity times mass).

Another significant feature of Quantum Theory is Wolfgang Pauli's 'Exclusion Principle', obeyed by all matter particles with fractional spin but not by force-particles such as a photon with integer (one-axis) spin. This means in simple terms (ref: Stephen Hawking's 'A Brief History of Time') that similar particles cannot both have the same position and velocity, within the limits given by the Uncertainty Principle (if they come close they must have different velocities and wouldn't be there very long). Quarks and electrons could never form atoms without the Exclusion Principle, but its cause was blamed on the Uncertainty Principle that Einstein strongly disagreed with. His Special Relativity theory forbids anything to exceed the velocity of light, but quantum theory predicts there is instant 'entanglement' between 'connected' particles. This was modified later to state that 'information' could not be transmitted faster than light.

Clearly, the disagreement was about 'determinism'. Einstein considered Quantum Theory incomplete because he believed that what controlled particle behaviour (i.e. quantum jumps, etc.) was not fundamental uncertainty but some undiscovered process. He may have been correct, but his concept of particles and fields probably needed refinement. An underlying reason for 'uncertainty' is that experiments can only sense the discrete energy of photons that have interacted, but not before!

Electro-magnetic fields and Photons
All particles traveling at less than lightspeed
have mass

Two small objects that are some distance apart (e.g.: hanging from strings) can exert electro-static forces on each other by exchanging photons. They usually have an equal number of opposite and alike charged particles and then all forces cancel, except for tiny gravity! If there is a surplus or a deficiency of 'free' electrons in either or both, (+ or -) electrostatic forces will overpower gravity. Although the 'Big Bang' particle formation process assured equal quantities of electrons and protons in the Universe, the structure of many materials allows (free) electrons in an atom's outer orbit to wander under the influence of external electro-magnetic fields. The resulting electrical unbalances are the source of natural electric phenomena such as electrostatics, lightning, etc. What keeps electrons orbiting around the nucleus of an atom at specific energy-levels is balanced attraction and repulsion from photon exchanges and orbital acceleration, plus the fact that energy only comes in discrete packages (quanta). We may call this the secret of our existence because our world is chemical, and interacting electrons captured by atomic nuclei form chemical structures.

You may remember from High School Physics that electro-magnetic fields consist of photons, traveling at the speed of light with a very large range of possible frequencies. A static electric field forms around electric charges but magnetic fields arise only between electric charges moving relative to each other. We were also told that scientists for hundreds of years argued whether or not photons were waves or particles but subsequently discovered that they are both. Marconi's transatlantic radio-transmissions early in the 20th century excited the public because it was magical, like X-rays.

Photons are different from all other particles. First, they are a concentration of energy traveling at the speed of light in a vacuum. We call it light, although there are many invisible radiation frequencies such as X-rays, gamma rays, infrared, microwaves, radar, and radio waves. All photons are electro-magnetic waves, with wavelengths ranging from .0000000000001 to 30000 cm. Secondly, they have no mass, only momentum. If they had mass they could not travel at the speed of light since that would require infinite energy! Thirdly, they spin about an axis in the direction of travel. The lack of clarity here is enough to scare off any high-school student, and some are not interested again until they retire.

The following is a direct quote by Albert Einstein in 1954, towards the end of his life: *"All these fifty years of conscious brooding have brought me no nearer to the answer to the question, 'What are light quanta?' Nowadays every Tom, Dick and Harry thinks he knows it, but he is mistaken. …I consider it quite possible that physics cannot be based on the field concept, i.e., on continuous structures. In that case, nothing remains of my entire castle in the air, gravitation theory included, (and of) the rest of modern physics."*

Today, quantum fields are assumed but they do not reveal the mystery of light quanta. Einstein may not have meant what he said; it seems illogical to contemplate matter consisting only of quanta in a quantum field and not connected somehow at a low level. We'd be back to billiard-ball physics, although may be I'm missing something. Quantum Theory and Relativity are correct for solving problems in Physics and Cosmology, but they do not yet relate to each other.

<u>Unification of Nature's Forces</u>
Particle-physics makes little sense without cyclic permanence

It is not the intention to give a detailed account of the present state of research in Physics. I am not qualified, nor would it improve anyone's understanding very much. There are hundreds of books available that can do a great job of confusing the uninitiated reader. Your attempts are doomed, unless you understand the mathematics, because none accommodate our accustomed need for mental images and analogies. My solution to this problem is the last major section of this chapter (entitled: 'A Vacuum Analogy'); it approaches things differently, although violating scientific rules and not blessed by anyone. However, my purpose is to question and illuminate, not to advance science! So far everything mentioned can be found in published information, with a bit of spice thrown in to befuddle dogma devotees.

The first thing I see, and have for a long time, is that true elementary particles with mass cannot possibly be static. This is nothing new; the Internet is full of people proposing that particles are cyclic entities. Although professionals agree that particles have a natural 'heartbeat', they don't seem to want to recognize this as playing a role in particle interactions or sustenance. Their equations predict everything well enough and that is all they care about, unless a very persuasive discrepancy should appear. Up to a point they are right, they should not be distracted by unessential speculation.

The reason particles with mass must have a cyclic existence is that mass, charge, and everything else make little sense otherwise. This argument holds no water with physicists because they are not concerned with visualizing what really goes on; mathematics is their reality and it puts a language barrier between us. Physics will ignore this concept completely and my apologies for being a nuisance are hereby offered. Many physicists continued Einstein's unsuccessful efforts to unify all forces of nature, ultimately concluding that he was wrong in his contempt for uncertainty in quantum mechanics but philosophically correct about other things (may be!). It was his stature that kept him respectable; otherwise he would have been laughed at, or ostracized like so many others. Einstein is on record for proposing that particles are localized pulses in a non-linear field, spreading out in space and moving as a stable unit. Other influential scientists expanded on this by pointing out that two particles really become one (with interacting reference systems), essentially making the entire Universe a single composite system.

Bohm and Determinism
David Bohm's version of quantum mechanics is
deterministic

Such a concept was explored by David Bohm, a very prominent American physicist who was blacklisted by the U.S. government after the McCarthy commission charged him with refusing to divulge information on communist activities of some colleagues (he refused on principle!). He died in 1992 in England where he settled after also working in South America and Israel. Early in his career he was at Princeton Univ. where he authored an authoritative textbook on quantum mechanics. He subsequently developed an alternate theory, equally compatible with experimental results and still quoted frequently. The physics establishment never embraced it, mainly because it did not predict anything new, but it was deterministic and lacked the probabilities that bothered Einstein. It supported Einstein by stating that the conventional interpretation was incomplete, but contradicted his views on non-locality ('entanglement', or 'instant action at a distance').

In his last book ('The Undivided Universe', 1993, Routledge), written with B.J. Hiley, Bohm postulates that particles do not have any permanent identity. It suggests that a particle cyclically forms, dissolves, and reforms at an extreme rate and never retains a stable form or field-location while dynamically interacting with other particles. Nothing has permanent form.

A VACUUM ANALOGY
Physical objects are not in space, but these objects are spatially extended. In this way the concept 'empty space' loses its meaning.
(RELATIVITY, 15th edition, 1952, Albert Einstein)

Four forces of Nature
Mathematicians hi-jacked Physics and any meaning is now irrelevant

This section (A Vacuum Analogy) is for those of us who are in awe of what physicists have achieved but still wonder about any meaning behind their mathematics! Many people believe Einstein was correct in assuming there has to be a cause for everything, even at the lowest level, and that the need for statistics only means that Quantum Mechanics is incomplete. Physics' celebrated 'Standard Model' recognizes four fundamental interactions of nature: Gravitation, Electro-magnetic, Strong, and Weak forces. They are reviewed here to explore their interdependence and to what extent knowledge of these interactions is empirical (probably all of it).

This is not a vital part of the book and it may be skipped over without affecting other topics. A theoretical relationship between the electro-magnetic and weak forces has been discovered (called electro-weak theory), but not yet for gravity or the strong force. The fundamental nature of force itself is also not well understood; physicists know it mathematically but seem disinterested in any physical understanding. That is not a problem directly, but it could be a reason for the lack of progress in some areas. The only way to proceed is testing of sensible analogies for speculated relationships.

Analogy-Model and Force
Particles are localized pulses and they move as a
stable unit

The analogy-model in this section describes assumptions that are consistent with experimental evidence reported by physicists and cosmologists. It assumes that what we call particles are quantised (discrete, or and non-spreading) states of energy and what we call the Vacuum is the continuous (non-discrete, spreading and chaotic) state of the same cyclic particles, dissolving and reforming at an extreme rate. Particles are then mutually sustaining and cyclically refracting (self-focusing) regions of concentrated density fluctuations (waves) originating from all other particles in the Universe and pervading the Vacuum. Such waves cannot be sensed (because they spread and are not discrete) except as gravity when their density is asymmetrically distributed near massive objects. Bohm was not the only prominent scientist to suggest cyclic particles. Quoting Einstein again: *"Particles are localized pulses in a non-linear field, spreading out in space and moving as a stable unit"*.

The nature of the Vacuum is the ultimate mystery; we think of it as 'something', and that is probably incorrect – it just 'is', if you can accept that. Actually, it should be imagined as nothing, although able to fluctuate positive or negative, but it works equally well (and is more popular) if assumed to have enormous energy-density with higher or lower density-fluctuations. Either way, they simply disperse at lightspeed to equalize Vacuum-density, except for discrete particles and radiation (photons) that have recurrent vortex refraction and therefore a local frame of reference.

It is inappropriate to present a complete analogy outline here, at least not in a book intended to be somewhat interesting. It is unlikely to fully represent reality and serves only as an indication of what actually may be going on in subatomic interactions. And why should I make a fool of myself by inviting accusations of arrogance for spouting unfounded opinions? Because in any scientific mystery there is room for intuition! There are other speculative proposals about the nature of gravity, electric charge, radiation, etc, but I like this one! Nevertheless, it is not science, only informed speculation!

The zoo of unstable particles is ignored completely; it is too complicated and they represent only intermediate stages in any case. They are not interesting for the same reason that a moving van is not a home. To put things in perspective, it is worth repeating that to my knowledge no one has any idea what is meant by the word 'energy', except as a transition between local temporary states (but of What?). Is this analogy just another crackpot idea, without proof? It probably is; except it does not claim reality, only usefulness as an analogy for Vacuum and matter interactions. You cannot be accused of deception when stating up-front that a concept is probably useful trickery.

Before proceeding, we should look at how long-range forces act on a single elementary (charged) particle. Photons change an absorbing particle's energy (and therefore its mass) when they interact with the particle's cyclically pulsating (vortex) waves. A vortex means that incoming and outgoing waves are not focused directly to a single point but tangentially to a spherical and diminishing (particle) region, and this accounts for particle spin about two axes (vortex spin about three axes is impossible!). The resulting interaction attracts or repels, depending on the charge relationship (+ or -) between a photon's emitting and absorbing particles.

You can visualize the entire process as a directional asymmetric modification of a particle's pulsating spherical waves. A particle's mean-density center then moves toward or away from the photon's source in the next cycle, and that describes force as a manifestation of cyclic displacement. The other long-range force is gravity which also distorts a particle's cyclic waves but always attracts any elementary particle proportional to its mass (energy) towards higher Vacuum energy-density. Gravity manifests General Relativity's 'curved space'. In contrast, when a photon is absorbed its energy is usually excess and immediately transferred by secondary photons into motion and/or heating (atomic vibration) of a solid object, liquid, or gas.

Attraction and repulsion from absorbed photons cancel when a light-beam strikes an object. Only their small momentum applies a tiny force called radiation pressure. The energy of photons not reflected is conserved primarily by object heating and subsequent radiation. A very large number of molecules are compressed when a cue strikes a billiard ball, shortening the atomic (standing wave-node) separations in both and triggering a repulsive exchange of millions or billions of uni-directional (virtual) photons between affected electrons. This restores the original (stable) atomic spacing, decelerates the cue, and accelerates the ball. The exact opposite occurs when a molecular structure is stretched. The foregoing roughly outlines my understanding of force at the particle level.

Gravity and Asymmetry of matter
Gravity is asymmetry of non-discrete Vacuum-energy around an object

The Vacuum has the ability to convey unlimited energy, unaffected by Pauli's 'exclusion principle'. However, it has no physical meaning for us although it is the foundation of everything we are, sense, or know (we cannot interact with it, except through gravity). It is like money in a cash-less society, where affluence is created temporarily out of nothing by borrowing. We conclude that gravity is caused by spatial asymmetry of matter surrounding a particle or macro object. Gravity acting on an object is proportional to its mass, confirming that it is not a surface-effect caused by directional waves such as photons since that would be proportional to any asymmetry of electric charges, and independent of mass.

Gravitational forces on a particle or object are in balance (cancel) when continuous (non-discrete) energy, received from everywhere in the Universe, is symmetrical. Such interacting (non-linear and chaotic) fluctuations have wavelengths in the order of the distance between quarks. Of course, they spread and reduce in density with the (inverse) square power of distance, which is why gravity is feeble and only noticeable near stars, planets, or other massive objects. Photons also spread and reduce similarly with distance, per unit volume, but the energy of an individual (discrete) photon stays the same over millions of light-years.

Gravity is best understood when it is considered entirely divorced from anything else we are familiar with. Greater energy-density of non-discrete Vacuum-fluctuations in some direction attempts to shift the mean (density) center of all elementary particles slightly towards it with every cycle. Therefore, spherical asymmetry of matter surrounding a particle or object is the cause of gravity (insensitive to charge, and always attracting proportional to mass).

The location where a particle reforms is determined by where it was in its previous cycle, but modified by accelerations due to electromagnetic, gravitational, and other asymmetries. The term 'location' refers to the particle's ref. frame in its previous cycle. It is important to keep in mind that we don't understand anything about the Vacuum. It has no structure and no mass; therefore its density-fluctuations do not resonate between (+/-) extremes, they only propagate and spread at lightspeed. This is the reason the 'Big Bang' formed discrete and charged (positive and negative) particles. Spherical pulse-waves of such particles cyclically disperse and refract (bend or guide) compatible Vacuum fluctuations towards the new particle-center in the same way that a lens refracts (focuses) photons. There exists a balance between the energy locked up in all particles and the chaotic fluctuations in the Vacuum; real particles will always be restored unless an anti-particle annihilates it. Gravity always attracts because (+ and -) fluctuations coming from a nearby massive object are spatially more energetic and distort restoring pulses to always accelerate a (+ or -) particle's mean-center towards the massive object.

Assuming overall asymmetry, all incoming fluctuations are refracted (curved) to exactly restore the original particle in the same location, after which they spherically expand in the particle's next pulse-cycle without loss of energy. This contrasts with discrete photon-waves which surrender all their energy to the particle, at least temporarily. In essence, it reduces all matter to self-repeating patterns of continuous interacting flows of energy-fluctuations (waves) that never persist in any form, somewhat similar to the proposed 'holomovement' by David Bohm, 30 years ago. The foregoing does not change the modern concept of gravity; it is simply an analogy to physically explain General Relativity's 'curvature of space and time'. However, it does propose that gravity is transmitted by continuous waves and therefore 'graviton' particles may be fictitious or virtual at best!

The 'speed of gravity' controversy is clarified by recognizing that photons have a ref. frame, whereas gravity does not (photons are discrete but fluctuations in a continuous Vacuum are not). Gravity acts directly (but not instantly) because Special Relativity implies that waves in the Vacuum have no ref. frame. Unfortunately, there may be other secondary (discrete) sources of asymmetry with gravity-like characteristics. One example is the familiar 'Lamb Shift' where polarization of a pair of virtual (non-permanent) particles, created in a single event close to any real particle, always aligns them to attract the real particle towards the source of the initiating photons. There will also be a minor 'shadow-effect', because energy-loss from acceleration due to gravity should reduce the gravity field between three aligned cosmic bodies. This suggests that gravity anomalies detected during solar eclipses (e.g. Maurice Allais, 1954) may not be a red herring after all.

Photons and Charge
Two pulsating electrons exchange virtual
photons when interacting

Like gravity, this subject is a perfect example of the kind of mathematical wisdom that mystifies non-professionals. The carrier of electro-magnetic force is the photon, a wave/particle that disperses and equalizes energy throughout the Universe. Photons are directional (discrete and non-spreading) waves that propagate at lightspeed in the Vacuum with a total energy proportional to their frequency. This makes little sense, unless you accept that a photon is not a simple bullet or a spreading wave of energy, and that the Vacuum in which it propagates is nominally at an average ground state of density but conveys photon fluctuations. Vacuum fluctuations can be spherical and non-discrete (spreading and undetectable), or directional and discrete (non-spreading) like photons.

Experimental data shows that photons have no rest-mass (their energy is entirely kinetic) and have vortex spin with left or right-handed angular momentum but very little forward momentum. They have transverse waves that are usually aligned (polarized) in a radial direction. Don't bother to visualize all this; no one knows what a photon looks like! If you really need a picture, remember that charged particles emit photons as excess energy. Photon-fluctuations may be imagined as discrete (non-spreading) density-bulges, super-imposed on a particle's dispersing spherical wave, and retaining this particle's spin about its directional axis. A photon can only be detected after a particle absorbs it; they don't interact like electrons that will scatter when colliding with each other. Experiments confirm that a photon is a non-spreading wave, acting like a discrete particle when intercepted (and absorbed instantly since it has no mass!).

However, photons have a problem! If they only have frequency, polarity, and spin differences (but no charge), how does one emitted by a positron or proton differ from one emitted by an electron and attract instead of repel other electrons? This is typically explained using Heisenberg's 'Uncertainty Principle', where photons are emitted by each particle in the appropriate directions to shove them where they are supposed to go (pushed together or forced apart). Conventional theory holds that the direction of emission depends on the charges of the source and recipient. This smells like magic; how do they know that from a distance?

Pulsating cyclic particles don't need such magic; their spherical waves are either + or − relative to the average Vacuum-density ground-state, depending on their charge. The Vacuum has dual states, similar to compression or tension in a neutral spring, but without any inertia and damping because only particles and structures with discrete energy have mass and heat-loss. Therefore, Vacuum-fluctuations do not resonate, they only propagate at lightspeed and spread (disperse) in one state or the other (+ or -). Photons don't have any charge, but they seem imprinted with source-charge information.

This is probably the right time and place to provide some details about the analogy used here (I hope you are interested!). When two stationary particles are close to each other they are both surrounded by each other's dispersing waves which can interact (refract) only where their curvatures are compatible, along an intersecting line between both centers. This asymmetric interference is either destructive or constructive and causes particle-reformation in the next cycle to separate (repel) their mean-centers with the same charge or approach (attract) with dissimilar charge. As a reminder, neg. and pos. charge may be seen as a cyclic excess or lack of energy, resp. An observer will assume that photons are exchanged or emitted to repel or attract, with no overall loss of energy. This matches the conventional interpretation without magic, although the concept of virtual photons causing repulsion is replaced by the above analogy!

The first time a high-school physics teacher mentions electricity you will hear about charges in uncharged conductors separating when brought close to any charged object such as a negatively charged rubber rod or a glass rod with positive charge. Free charges in an uncharged object will move to one side or the other and always cause attraction towards any charged rod. You heard that this causes no change in overall object-charge, only a shift of free electrons. This is your introduction to electricity, and you may even proceed to become a physics teacher yourself without ever questioning photon-exchanges as the reason for it all. It is mathematically sound but it makes little physical sense, which is the subject of this tiresome analogy. Some additional information you probably don't want: free electrons are rare in insulators but common in conductors.

Two interacting particles always align their spin-vectors to achieve the lowest potential-energy, causing repulsion when both charges are the same and attraction when dissimilar. In the first case they must share available Vacuum-energy and in the second they fortify each other. The emission of two real photons are a consequence, but not the cause of attractive acceleration between dissimilar charged particles. They carry away resulting excess potential energy, with spin and polarity differences but without charge. They will give up this energy to a spin-compatible distant recipient (electron or proton), increasing its kinetic energy. Since two photons with opposing spin are created, each has a 50% chance of either attracting or repelling any distant absorbing particle in an uncharged object (because there are an equal quantity of pos. and neg. particles in the Universe).

Photons with a specific spin-direction will cause opposite reactions when absorbed by a matter or an anti-matter particle. This is intuitively difficult, although similar to $(+1)-(+1)=(+1)+(-1)$ since anti-matter is mirror-imaged in every respect. It answers the question how photons remember the charge of the particle they originated from: they don't remember! This is unimportant between uncharged objects because everything cancels and photons only carry energy away from one object to increase it somewhere else. However, charged objects emit and receive unequal numbers of left and right-hand spin photons, because free electron or ion spin-vectors align themselves pointing away from the object. Such a biased flow of photons received by an uncharged object energizes and aligns, but only detaches and displaces electrons (not ions!), causing unequal charges on either side of a conductor. It is therefore inequality of exchanged left and right-handed photons that is the source of attraction, repulsion, and electric-charge, not individual photons!

This describes photons, fundamental to the exchange of energy in our matter-Universe. It provides credibility for the concept of rapidly dissolving and reforming particles, assuming the above analogy has any merit. It is obvious from previous and following sections that electrostatic and gravity forces are different but do have a common origin, as do nuclear binding forces. It opens the door to a "Theory of Everything'!

Of course, there is much more to this story, although not all appropriate here. For instance, if you wonder why photons have a frequency, consider the following. Atoms in red-hot iron vibrate due to thermal energy and their electron-groups constantly bounce into each other, triggering reactions with their nuclei. It causes the emission of real photons (thermal radiation) into the Universe, with a corresponding loss in kinetic energy of the colliding atoms. Standing waves lock electrons in specific orbits (shells) around the nucleus, like a water-skier stuck in a wave-trough behind a ski-boat. They will temporarily switch ('jump') to higher shells after absorbing a photon, only to be forced back immediately, causing the emission of two real photons. The rate of quanta (frequency) generated by an atom depends on the kinetic energy it has lost. This is why light from white-hot iron has a higher frequency than red-hot iron! Photon-frequency is indicative of temperature and the chemical composition of the emitting source, by its spectral signature.

And another thing: cyclic electrons and protons don't need to emit virtual photons continuously to maintain electron-orbits (as per conventional theory); electrons reform in stable orbital paths at standing-wave nodes, due to the complex spherical pulse-patterns from the nucleus and neighboring electrons. No energy is lost, therefore no real photon emission, and virtual photons are needed only to balance the instantaneous energy books. Real photons are created in many different ways, but conservation of energy is always at the bottom of it.

One other mystery needs clarification: the spin-vectors of all free and most other electrons become aligned (directionally pushing each other) in a conductor with electron-current. Therefore, uni-directional vortex refraction of Vacuum-fluctuations causes rotating magnetic fields around a current-carrying conductor (not photons or magic!). From the well-known magnetic field right-hand rule around a conductor we can conclude (assuming the analogy is valid) that all electrons, and the photons originating from them, have right-handed spin. Left-handed photons then originate from the positively charged particles, although handedness can be switched. This is another of Nature's coin-flips and similar to the disappearance of all positrons into nuclei, instead of all electrons.

The energy of photons can create virtual positron/electron pairs. This is reminiscent of an old Chinese philosophy (Yin-Yang) that any force in nature is counteracted by an opposite and that one defines the other. An electron does not have more or less rest-energy than a positron; it has the same, but in the opposing energy state. An electron cannot be created without a positron and any remaining energy is conserved as kinetic energy, separating them. Virtual particle-pairs normally annihilate each other almost immediately, but they can be elevated into real particles if the initiating photons had sufficient energy to accelerate them far enough apart. Of course, positrons don't last long near electrons. Mutual annihilation then creates a burst of (photon) radiation, only to be used somewhere else to create again!

The foregoing suggests a simplified but plausible scenario of how matter and universes are created. Assuming that at the end of any universe's life all energy collects in a super black hole as radiation, then eventually its density rises high enough that virtual electrons and positrons become real (permanent) at a critical rate. As mentioned before, leptons (positrons and electrons) are unprotected from annihilation unless all of one are captured in nucleons and this happens only during the enormous kinetic energy and density in the initial phases of a "Big Bang". A significant fraction of the black hole's accumulated energy is then converted into real particles in a runaway 'Big Bang' and 'voila!', a new Universe is born. This may be the 'Creation'-process of matter!

Observing a photon in our, and then in the photon's, ref. frame is similar to the misalignment between the Earth and Moon's gravity and light. We would see a curved lateral wave, but it is perfectly straight as far as the photon is concerned. For us, it must be curved, otherwise the outlying part of the wave exceeds the speed of light, but time stands still for a photon. This is a fundamental observation that explains many mysteries in Quantum Mechanics, including how a single photon can interfere with itself when passing through two slits at the same time.

Photons are clearly non-spreading waves with a fuzzy (probability) center, usually polarized in a random radial direction, and the 'Standard Model' QED theory confirms this. However, Einstein's photoelectric theory proved that they are also particle-like (discrete) after interacting with charged particles. Regardless how many photons are absorbed, only those above a minimum energy level (frequency) can knock electrons out of certain metals. This makes sense with the following analogy: throw a thousand little pebbles at a coconut in a tree and it won't budge; combine them in a bag and you'll knock it loose! Wave/particle duality is well verified and an accepted fact.

My apologies for this long and speculative section on photons (the uncertainty of it all made me do it and, after all, this book is about uncertainty!). Assumptions about gravity and photons are probably incorrect, at least in detail, but there is no easy approach to these quandaries. Reality is bound to be complicated; maybe too complicated to be an accident! We may not fully agree with him, but we can understand Fred Hoyle's comment: - *the facts suggests that a super-intellect has monkeyed with physics, as well as with chemistry and biology, and that there are no blind forces worth speaking about in nature.* – It does seem outrageous if only chance is responsible; and although the Vacuum is subject to rigid rules, its potential appears unlimited.

Entanglement and Weirdness
Separation and time are illusory perceptions,
caused by relativity

We have now reached a subject I dread: what is meant by quantum 'entanglement', also known as 'non-locality' or 'instant action at a distance'. It is obvious from these names that it is weird, and you are right; it is exceptionally strange. It requires revision of fundamental concepts about reality we formed in early childhood. The initial reaction will be disbelieve; you'll probably question my sanity or assume that something else is going on and has been misinterpreted, but you'd be wrong (especially about my sanity)! Particle entanglement has been experimentally and theoretically confirmed and is a scientifically accepted fact (e.g. Alain Aspect, Paris, 1982), but it is clearly inconsistent with our reality.

When two photons are created in a single event, and repelled in opposite directions, they stay connected even though they may be light years apart. Change the polarity or spin of one and the other also changes instantly! It seems impossible and Einstein's Special Relativity Theory says that as well, but it is wrong! Without further embellishment here is my opinion for what it's worth, although there are as many opinions as there are people who have thought about it. We obviously look at it from the perspective of our ref. frame. Now look at it from a photon's point of view. Traveling at the speed of light, relativity dictates that its time stands still relative to our ref. frame, and the same goes for the other photon. Our reality for both photons is then that they are still connected, since no time has passed for them and separation is not even feasible. OK, nobody else wants to accept that either, but the test-results don't lie!

So we have a very weird Universe where perceived distances depend on the relative velocity of frames of reference. Particles can cash in their mass in one single reaction and travel as two photons in opposite directions at light velocity, while remaining instantly connected. The only way to visualize this from our perspective is to assume that a virtual single frame of reference in the form of an instant (tunneling) channel is allowed, something like a true void (a worm hole?) within the Vacuum with an interaction speed that appears to us infinitely higher than lightspeed. Before you dream up all sorts of ways to utilize this phenomena, and on the off-chance you believe any of it, there is a catch! Our reality consists of discrete energy (matter), and it is limited by the speed of light, faster is impossible. In other words, entanglement applies to photon-waves, not a people world! It could be useful in the future for quantum teleportation and cryptography, but only at lightspeed!

Considering the size of the Universe, being told that there can be a (near) instant connection between opposite ends is mind-boggling. The closest you can come is to assume that our Universe is inside a black hole, a dimensionless and timeless singularity depending on how you want to examine it. It makes some sense, if you can twist your mind in the right direction. For instance, it clarifies what exists outside of our Universe - absolutely nothing we relate to, because time and distance exists only for particles with mass! It also answers the question raised at the beginning of this book - is what we experience real or only a projection? It is real, but our sense of separation and of certainty is illusory. Most people, at least 99 out of 100, refuse to believe any of this; it is too uncomfortable and distorts their sense of reality and security. They don't want to think about it.

I am sorry for this crude and amateurish picture, but it exemplifies what this book is all about. There are physicists who explain entanglement by using the analogy of backward time travel, but that is even more bizarre. You can continue to assume that (in your reality) enough time will elapse to swing a bat when a pitcher throws a baseball. The reason is that you, the bat, the ball, the pitcher, and Earth consist of localized bits of energy called electrons, atoms and molecules, and they are all part of macro-structures with nearly identical ref. frames; all originated in close proximity in the Big Bang. This is not the case for energy in the form of electro-magnetic radiation or gravity, because then space and time is very different or non-existent. Another way to twist reality is by imagining that things you know to be solid (e.g. a coin) are actually a maelstrom of discrete energy-fluctuations in a Vacuum that is continuous (undivided), imperceptible, and without structure of any sort. Nobody really understands this concept of Vacuum-energy; it is like discussing nuclear science in the Middle Ages. And take my word for this much: it is a highly unpleasant subject before going to sleep.

Strong and Weak Forces
Positrons lost a coin-toss to electrons and ended
up captive

It is better not to elaborate too much on QCD (Quantum Chromo Dynamics) or the 'Strong Force' that acts between nucleons or quarks, because it is a minefield. However, I do have some speculative opinions, and might as well tell you what they are. It is possible that nucleons (protons and neutrons) were formed out of primeval free and ultra-energetic electrons and positrons when the 'Big Bang' began and the local Vacuum was ultra dense. Positrons lost a coin-toss to electrons and they all ended up captive (forming a matter Universe, instead of anti-matter). They, and a large fraction of electrons, combined as quarks in nucleons while retaining their relativistic velocities (and large mass). This was fortunate because it protected them and all electrons from annihilation. Leftover free electrons lost a lot of thermal kinetic energy (and mass) by powering the Big Bang accelerated expansion, until they could be captured in proton orbits. Why does the 'Strong Interaction' lock quarks in such an unbreakable embrace in protons and neutrons? There may be a number of reasons:

1. Pulsating energetic particles (quarks) at such a close range cause standing waves and nodes in an enormously intricate wave-field where both positive and negative charges are dynamically phase-locked in place (with a rotating frame of reference for each interacting pair). Within limits the three quarks are relatively free inside a nucleus but unable to escape or approach. This is called 'asymptotic freedom'.

2. 'Strong Force' interaction is an opposing energy-state attraction, strongly enhanced at such short distances within a small group of particles. With three quarks, the odd charge inside a nucleon is always located somewhere between the other two and therefore two pairs attract each other but only one pair repels. In a cluster of four

(2+2), each particle is attracted by two others but repelled by only one, and so on.

3. The interacting force that binds quarks is strong enough that the energy needed to separate them in a collision is sufficient to create new quarks, and that prevents ever seeing individual quarks.

Present theory says that in a neutron two (-1/3 electron-charge) 'down-quarks' electrically neutralize a single (+2/3) 'up-quark'. A proton with two 'up-quarks' and a single 'down-quark' then has a charge of +1, which is neutralized by its captive electron. Neutrons are unstable outside of a nucleus, and will decay into a proton. The additional energy of a neutron's less efficient structure is liberated by casting off an electron and a neutrino. A quark's rest-mass energy (at zero velocity) should be the same as that of an electron or positron, but fractional charges probably arise because that is the energy left over for external interactions. The remainder is binding energy in nucleus gluon exchanges. A neutron is electrically neutral and does not interact with electrons. It therefore can pass through matter, which makes fission of heavy nuclei practical. In a rough sense, all of these assumptions should be compatible with QCD.

This leads to the question what Neutrinos really are, since the decay of a neutron into a proton by the so-called "weak-interaction" produces a neutrino in addition to an electron. In fact, decay of many unstable particles often results in emission of a neutrino of one kind or another. This section is already getting much too long and some shortcuts are needed after all the heretical speculations. A neutrino may be a longitudinal energy pulse, not transverse like photons. It is emitted due to internal orbital realignment, also ejecting an electron when a neutron decays into a more stable proton configuration. The macro equivalent is hitting a guitar headstock lengthwise, causing longitudinal waves in the strings.

Unlike photons, neutrinos penetrate right through the Earth and millions pass through your body every second. When they line up just right, pure chance causes very infrequent interceptions by particles. The interesting thing about neutrinos is that they do have an extremely small mass (only recently confirmed). Of course, the longitudinal wave is real for the neutrino but foreshortened entirely in our frame of reference! Even with a tiny mass they cannot be traveling at lightspeed, only very close to it, and they will infrequently accelerate intercepting particles backward, like negative gravity. It is a good example of a tiny bullet of energy, accelerated to nearly the maximum velocity allowed in the Vacuum, with kinetic energy equivalent to $E=m*c^2$. It should be a leading candidate for the mysterious "dark energy" supposedly accelerating the expansion of the Universe.

Mass and Inertia
'Higgs' particles are redundant with cyclically
pulsating particles

Next we must consider mass and inertia, incompatible with fluctuations or waves at lightspeed in a continuous Vacuum and applicable only to discrete and cyclic parcels of energy. Inertia and mass are more or less the same thing, like a loaf of bread and the amount of money you pay for it (mass is the quantitative measure of inertia). Inertia arises when atomic electrons of two bodies interact and accelerate (repel) each other. It derives naturally from the assumption that discrete particles are cyclic pulsators. Cyclic particles establish time and space in a Vacuum that is otherwise timeless and dimensionless. It requires external energy to relocate a particle's frame of reference and that gives particles mass!

A particle's mass relative to some ref. frame increases (or decreases) by merging and absorbing (or emitting) photon-energy. This is not magically caused by acceleration (or deceleration) and it confirms the fundamental equality of mass and energy! The concept of pulsating particles eliminates a great deal of mystery in nature, especially gravity. Inertia is not applicable to the Vacuum itself, but the force of gravity on an object is proportional to its mass. Any object, from a single particle to a cannonball, is accelerated at exactly the same rate in a gravity-field, discovered by Galileo. Mass has been attributed to many weird causes but, without a discrete rest-energy, mass cannot exist. And it is difficult to see why a hypothetical "Higgs" field and particles are needed to provide the Vacuum with a fluctuation-texture that gives everything mass.

How can you determine the mass of an object? There are two ways: you either accelerate it and measure the force required, or you weigh it in a gravity field. Both methods are basically the same because gravity acting on an object is equivalent to accelerating it. In either case you need a calibrated mass to compare against. Throw a rock and the resisting force is inertia, caused by slightly compressing molecules in the rock, and many particle receiving uni-directional (virtual) photons from each other. Your muscles must supply more energy when you throw

122

faster or when the rock is bigger (more massive). It is mind-boggling to consider the number of photons required (in your body and in the rock)! I probably need to wait a few more years before explaining this to my grandsons; they are still into the mechanics of rock throwing rather than the physics. They may become interested eventually, but my wife and daughters wouldn't listen to it for more than 5 seconds. Even so, I should explain it to my granddaughters as well, or they'll accuse me of gender stereotyping!

Cosmological Uncertainties

'Big Bang' extreme density made virtual particles real

We can quickly revisit two recent cosmological puzzles: 'dark matter' and 'dark-energy'. Every current science magazine will have some article about these mysterious subjects. Dark matter is believed responsible for the gravitational observation that galaxies appear to have much more mass than what is visible. Exotic forms of matter are blamed, but it could be many other things. A long shot speculative reason for the dark matter problem may be charge asymmetry from polarized virtual electron/positron pairs ('Lamb Shift'), triggered by energetic photons from galactic sources and modifying long range Newtonian gravitation. Another related mystery is the so-called "Pioneer anomaly". Reduced density of free hydrogen inside of our solar system, due to its rotation, could (among many other possibilities) account for this unexplained deceleration experienced by two (early) 'Pioneer' spacecraft. Dark matter is a subject that may yield major insight into new physics in the near future, when more accurate satellites become operational.

'Dark energy' is blamed for the observed frequency-shift of very distant supernovae, indicating that expansion of the Universe is speeding up. The most common hypothesis is that 'zero-point' (ground-state) energy of the Vacuum is responsible. Without giving it much thought, it seems reasonable that expansion of the Universe is accelerating because we know galaxy-clusters are separating and gravity is losing the battle (a leaky Universe!). Neutrinos could also be the culprit because there are so many of them and, even with minimal mass, their momentum must apply some outward pressure as described earlier. Conversely, accelerated expansion may be only one phase of a closed cycle, since everything else cycles in nature!

The most mysterious objects in the Cosmos are Black Holes and their cousins, Neutron Stars. The latter can actually be seen, because they emit photons while rotating at an enormous rate. Black holes can be detected only by their gravitational influence on other stars; light cannot escape their gravity and (nearly) all discrete matter and radiation is assumed to collapse into a singularity. They represent the final stage in a battle where gravity triumphs over electro-magnetism, at least at a scale less than galaxy-clusters. If dark-energy defeats gravity in the Universe it will be a one-off event! Stars like the Sun burn out and eventually become white dwarfs; slightly larger and they become neutron stars. Neutron stars consist primarily of neutrons after a Supernova ex/implosion jams most electrons and protons together into neutrons. Bigger stars (more than five times the Sun) become black holes at the end of their nuclear-fusion life, when gravity crunches everything into oblivion.

How does this speculative 'Vacuum Analogy' relate to the 'Big Bang' scenario envisioned by Cosmologists? There are many possible scenarios. First there was some initial event, because logic dictates there had to be one. This has all the ingredients of what many people regard as a Supernatural miracle. It must have been triggered by some instability in a perfectly smooth something, balanced on a knife's edge. Of course, the obvious question still is - what is energy's conveyance, what made it or where did it come from? Speculating on subsequent scenarios is easier and it may have been a giant primeval 'black hole'; and hopefully the Universe will always remain inside its event- horizon! A lack of external gravity at its low cycle will gradually compress it more and more. The end-result may be that virtual electrons and positrons in its center receive enough energy to become real at an exponential rate, and a new 'Big Bang' is born! But there are other possibilities, including a unique one-off occurrence with energy now dispersing.

Magical Universe, David Bohm
Thank God for who and what we are, set free by
our imperfections

Our Universe is magical; there is no other way to describe it. And I understand why Newton looked for magic, alchemy and religion to explain its secrets 300 years ago! What we perceive as structures of static substances is in reality a seething flow of sub-atomic fluctuations (waves), constantly forming, dissolving, and reforming discrete particles. The unimaginable magnitude of this is entirely beyond our comprehension. Without the weird (discrete energy parcels) aspects of quantum mechanics all particles would be unstable and our Universe could not exist. As mentioned before, a bridge can stand for centuries and it does not slump into the river (or get tired, like us). Molecules are perfectly stable unless excessive energy beyond a critical level is supplied, like a ship hitting the bridge.

At our macro level we see buildings made of stone, brick, wood or steel, etc, but none of that is actually solid or inert. If we (in our imagination) could magnify their composition millions of times more than we are now capable of you would see a chaotic four-dimensional flux of Vacuum density-gradients, following orderly principles that sustain our world, including us. And if you are intrigued, but not impressed by my amateurish attempt at determinism, please check out David Bohm's theories (especially his 'holomovement' concept). He would scorn much of my writing and call it unscientific, but in the end only conceptual meaning is important. A few Internet clicks will explain his authoritative version far better, although most of the mathematics were simply too complex for me!

Bohm's holomovement means that physical separation between particles and objects is only an illusion, caused by the reality of our local frames of reference. In the reality of the Vacuum there are only interactions with no duration and no separation. In the mean time, I am satisfied to conclude that we are not controlled by a single autonomous process of predictable interactions. Following the principle of 'Yin-Yang' in ancient Chinese philosophy the interactions of discrete energy (matter) are made possible by non-discrete (Vacuum) energy, and vice-versa. This opens the door for conscious semi-logical beings like us to make decisions based on 'free will'.

Although we all accept free will as a foregone conclusion, it is nice to have it confirmed. Looking at it another way, our logical mind perceives and chooses between what we remember as reality and imagination (illusion). Suggested previously was that 'free will' originates from conflicts between logic and instinct, and from Heisenberg's "Uncertainty Principle". In the absence of uncertainty we would be slaves of unalterable logic; the reason any functional 'Artificial Intelligence' shall also require 'uncertainty', or some degree of instinct or intuition. And thank God for who and what we are; set free by our imperfections!

It is questionable that this 'Vacuum Analogy' section has any scientific significance; it is conjecture and probably mostly wrong. But it outlines what may be possible beyond what we know. It strongly suggests that everything must depend on everything else. Einstein discovered that time and space depend on local energy-concentrations that vary in time and space, etc. Nothing is constant, although it appears that way from our transient point of view. Science's so-called constants of Nature are local attributes, varying when underlying ambient factors do. There is no orderly little house in an orderly world; we exist because the possibilities are infinite!

CHAPTER IV -SOCIAL SIGNIFICANCE OF SCIENCE

STRUCTURES AND ELECTRONS
Energy Conservation laws state that energy cannot be created or destroyed.

Chemistry and Electricity
Everything we sense and know is caused by the
exchange of photons

It may not be immediately obvious that chemistry, the science of the composition, structure, and properties of molecular matter, depends entirely on energy exchanged by photons between electrons and the nuclei they are prisoners of. Without that, chemistry and the enormous variety of chemical substances would be non-existent. At least 30 million chemical compounds (molecules or crystals) have been identified, formed by chemical reactions (combinations) originating from about 80 non-radioactive chemical elements. The atoms of one specific chemical element are all basically identical, with the same number of protons. A majority of chemicals are now polymers and plastics, used for consumer goods, accounting for 80% of the output of chemical industries today.

Everyone has a concept of what electricity is, and we are all mostly wrong. It does not matter because designers of electric equipment know the relationship between design parameters very precisely, based on experimental data. You will later encounter a section called "Electric and Magnetic Fields" that reads like a mini-textbook. Like all other sections, it is easily skipped over, but it emphasizes the importance of particle/photon interaction in electrical apparatus. There were a large number of historical contributors, but none more important than two 19th century British researchers. The electro-magnetic field was first fully described mathematically by James Clerk Maxwell (1831 –1879), a Scottish mathematician and theorist. He based his theories on the test-results of many earlier experimenters and particularly Michael Faraday (1791–1867), the English chemist and physicist, who turned out to be exactly the right person needed at that particular time in history.

Faraday and the Royal Institution
Initial progress in electro-magnetism depended
solely on analogies

The son of a poor blacksmith, Faraday had minimal education but he could read, and that was enough! He elevated himself gradually into a world class experimental chemist and physicist, initially helped along by his access to books as an apprentice bookbinder. This was at a time when science in Great Britain was conducted by 'gentlemen', and not by the son of a blacksmith. He bound all his notes taken during local scientific lectures and, after noticing it in the bookshop, a few people arranged for Faraday to attend formal lectures at the Royal Institution in London. As a courtesy he sent the famous chemist Humphry Davy the bound notes of his lectures and, in time, Davy offered him a job as an assistant in the laboratory at the Royal Institution. Davy treated him as a manual helper, had him carry bags when traveling, etc. But other people recognized his natural genius and he advanced rapidly, against the wishes of a jealous Davie. Preoccupied at first with chemistry, he made several important discoveries and became laboratory director in 1825 at the very young age of 34.

The importance of electricity in chemical processes led him to investigate electrical and magnetic phenomena and that made him very famous eventually. He contributed to a large number of early applications of electro-magnetism and electro-chemistry during his lifetime. This included the invention of cathode rays (electrons), an early indication of the quantum nature of matter. Equipped with only basic theoretical knowledge, he had an extraordinary ability to pictorially imagine the puzzling aspects of what eventually would be called electro-magnetism. Despite the ridicule of virtually all other scientists it was his prediction and use of the abstract concept of electric and magnetic 'lines of force' that made him stand out among his lesser but more learned peers. He showed in 1831 that moving a current carrying wire creates a magnetic field and that moving a magnet near a conductor creates an electric voltage. Called induction, it is the foundation of electro-magnetic technology. His imaginary lines of force are what Einstein meant, when he said:

"Space-time is not necessarily something to which one can ascribe a separate existence, independent of the actual objects of physical reality. Physical objects are not in space, but these objects are spatially extended. In this way the concept 'empty space' loses its meaning."

Faraday did not worry about what caused it, he simply observed and relied on whatever analogy was convenient to explain it; his insight was intuitive, not theoretical. His practical approach bypassed many scientific roadblocks and it generated phenomenal progress. He eventually construed light as 'vibrating lines of force' but, lacking the mathematical skills to elaborate, he received little credit for this perceptiveness. It is interesting that in 1839 he was 50 years ahead by rejecting the need for a stationary 'ether', when much later even Maxwell could not yet accept that. The meaning of his 'lines of force' is identical to the modern concept of 'electro-magnetic fields'. He was modest and very religious, turning down a knighthood and (twice) the presidency of the Royal Society. He refused to participate in the development of chemical weapons for the Crimean War. Faraday died in 1867; but his impact on civilization is evident each time electricity is put to use and it makes him an authentic hero in history!

Maxwell and Electro-magnetism
Maxwell's equations are at the ultimate of classical science

It was Maxwell who picked up the loose ends from Faraday and many other experimenters, developing it into a mathematically consistent theory of electro-magnetism, thereby unifying several separate electric and magnetic theories. Faraday lived long enough to see his 'vibrating lines of force' concept confirmed as how our Universe operates, by transferring energy. Forty years younger, Maxwell had little direct contact with Faraday. He was born (1831) into a wealthy family and received a solid education, excelling in mathematics. Starting in 1856 he was appointed professor of 'Natural Philosophy' at a series of colleges and it was during this period that he became well known for his invention of color photography. He demonstrated that three (black and white) photographic plates of a subject, using red, blue, and green filters, could be recombined into a color projection by using three projectors equipped with the same filters. When his father died he returned to his family's estate in Scotland in 1865. Here he worked for 8 years on the details of his now famous four partial differential equations that define the connection between light and electro-magnetism. He was appointed head of the Cavendish Laboratory at Cambridge University in 1874, but died of cancer five years later.

Maxwell's impact on Physics is difficult to put into a few words; his equations are fundamental to all physical processes and they were only modified in the 20th century by Quantum and Relativity Theories to accommodate ultra-extreme conditions of size, velocity and energy density. It did not matter that his equations used an imaginary ether, it worked (under ordinary circumstances) and allowed the basic calculation of all electric and magnetic interactions. He established that when this fictitious ether's elastic properties were adjusted to fit the (electric and magnetic) experimental results, such forces were transmitted at the speed of light. This established that light was also electro-magnetic, although it was not fully accepted until his death in 1879. Einstein eventually dispensed with a stationary ether in 1905 with his Theory of Special Relativity.

Electric and Magnetic fields
Photons fill the Vacuum and E.M. fields create
order out of chaos

Electro-magnetism (E.M.) is force by the exchange of photons on a scale detectable at (our) macro level, called an E.M. field. Although this is not intended as a physics lesson, the subject is perplexing enough that ignoring it perpetuates misunderstanding. Since kids tend to select mysterious electricity as a subject for science-projects, it enhances your reputation if you know something about it (even when scientists laugh at you!). Conventional explanations are based on a flow of (negatively charged) electrons and it provides the necessary equations, but you'll never really understand it! Electro-magnetic fields are generally considered streams of uni-directional photons on the scale of objects, rather than particles. Electric and magnetic fields are interdependent, acting perpendicular to each other. The electric field is created by stationary charges and the magnetic field by moving charges.

In any concentration of free electrons (e.g. in a battery or a capacitor) photons are emitted and absorbed in all directions, squeezing out weakly linked electrons and creating a surrounding static electric field. Mutual repulsion seeks to reduce the charge-density of such fields and any suitable conductive path provides an escape (an electron-current) towards lower charge-density. Pushed by photons, the spin-axes of all electrons in an electric current become aligned. Vacuum-fluctuations, refracted to cyclically restore each electron, then curve uniformly towards the electron's center from everywhere, strongest at its spin-equator and zero at its poles. This forms a right-handed rotating magnetic field around the conductor.

Two parallel conductors with current flowing in the same direction experience attraction because the field between them is stronger. The opposite happens with opposing current flow. Until conductor resistance stabilizes the current, accelerating electrons create a varying magnetic field that will accelerate electrons in any nearby (parallel) conductor, inducing an electron-current in the opposite direction. Resistance to current in a conductor arises out of chaotic but symmetrical (photon) thermal interactions. Current in the wire of a coil is limited by self-induction because increasing magnetic flux induces an opposing current in the same wire.

Therefore, what is called electro-magnetic induction is a varying electric field that causes a varying magnetic field, causing a varying electric field, etc. A steady electron flow (direct current or D.C.) in a conductor does not induce current in nearby conductors because everything remains spatially symmetrical (balanced). Alternating current (A.C.) induces electron-current in a secondary circuit and is therefore more useful for electric motors, transformers, and many other electrical applications. Permanent (ferromagnetic) magnets have the collective spin-vectors of many atoms locked in a certain direction during manufacture by heating and rapid cooling. Each atom (or group) is then a tiny magnet contributing to the overall magnetic field.

Developments in electrical engineering contributed more than almost anything else to the technical revolution that began in the 19th century. One well-known application of induction (at long range) by electric and magnetic fields is the transmission of radio and television signals through the Vacuum. Free electrons oscillate back and forth in the antenna of a radio-transmitter, broadcasting photons through the Vacuum. These photons cause electrons in a receiver's antenna to oscillate in response and any superimposed signal is then discernible when it is tuned to the same carrier frequency.

An avalanche of refinement in 'electronics' was initiated in the late 1950's by the invention of the Integrated Circuit, also called silicon-chip, microcircuit, or 'I.C.' for short. It grew out of the earlier invention of 'solid-state semiconductors', which replaced vacuum tubes in radios, etc. A semi-conductor allows passage or non-passage of electron-current under the control of a secondary electric circuit. There is no need to emphasize the importance of the I.C. revolution; most modern households are saturated with analog and digital applications such as cell-phones, computers, etc. Miniaturization has been mind-boggling and it is difficult to compare the historic influence; it is unlike anything that happened before. It came out of nowhere and its impact on society was like the combined inventions of steam, fossil fuel, and electrical power.

Chemical elements and Supernovae
Charge and gravity duality is the cause of Nature's complexity

There are 94 natural chemical elements found on Earth, and 92 of them were created inside stars. Out of these, the 12 heaviest (Bismuth and above) suffer from radioactive decay because they are inherently unstable. Radioactivity consists mostly of alpha, beta, and gamma radiation (helium nuclei, electrons, and photons, resp.). The Sun has been stable in its radiation output for at least 600 million years and will remain stable for another 6 billion years. If its fusion rate of hydrogen into helium were to increase, the Sun will expand and its core temperature decrease, resulting in less fusion. This is why the Sun's radiation on Earth always cycles about an average and stable value, although our atmosphere can interfere. This is called negative feedback, and it is one example of the many fortunate coincidences allowing life on Earth to proceed. Fusion of progressively heavier nuclei takes place in a series of heating and cooling cycles within all stars, until all helium is depleted and converted to carbon, after which a star like our Sun becomes a red super-giant. It will slowly cool into a white dwarf and then, after many billions of years, into a dark dwarf.

Stars more than twice as big as the Sun suffer a different fate, and they have played a crucial role in the composition of Earth. Their core is much hotter, and fusion proceeds much faster than in the Sun. In contrast to the carbon core in the Sun's future, the eventual core of massive stars is iron which cannot be fused into heavier elements because that would require more energy than additional fusion can produce (the protons and neutrons of iron are packed as tightly as is physically possible). When the fusion of surrounding lighter elements is exhausted this iron core gets much hotter as it contracts. Electron repulsion can no longer overcome the crushing self-gravitation and a Supernova occurs, after a (one second) fatal collapse. The potential-energy of the former star's gravity is so large that during the contraction many nuclei (including iron) are fused into much heavier nuclei. Up to 90% of its mass is then blown into space after the violent rebound. Gravity has formed our Solar system out of that kind of primordial debris. Rotational centrifuging left only the lightest elements (hydrogen and helium) in its center, out of which the Sun was formed eventually. The Sun is relatively small and therefore slow burning and still young; its original 27% (Big Bang) helium content has increased to 62% and the rest is unfused hydrogen. Many generations of fast-burning massive stars formed and blew up before the Solar System formed and that is how Earth inherited such a bonanza in heavier elements.

<u>Chemistry and Carbon</u>
The most important element on Earth for
biology is carbon

All structural matter and therefore the science of Chemistry depends on balancing or matching of energy-levels between electrons orbiting the nuclei of atoms, particularly those in outer shells. It is assumed that standing waves between particles are responsible for the separation, location, and binding of electrons in the minimum energy positions of wave-nodes. This determines the number of electrons allowed per shell and causes the mysterious instant jumps from one energy level to another. It is not mysterious if you accept that electrons and quarks pulsate cyclically and that their wave-nodes are super-positions; any electron that jumps an energy level simply terminates in one node and reforms in another. This approach (more or less) was advocated by Erwin Schrödinger, an Austrian physicist and one of the initiators of Quantum Mechanics. Real photons are not exchanged between orbiting electrons and nuclei (when there is no change in energy-level). Dispersing pulses of these particles refract converging Vacuum-fluctuations into new particle-vortexes and it is their wave-nodes that form chemical structures, from air to elephants and mountains, and without which the Universe would be one huge radiation fluctuation spreading forever in the Vacuum.

The most important element for life on Earth is carbon; it makes up 18 % of our body, and only oxygen exceeds that. Since it is an abundant element throughout the Universe, it is also the likely basis for extraterrestial life, if it exists. It has six electrons but only four in the outer shell are responsible for its chemical usefulness. Carbon is unique for thousands of compounds needed in life processes, when combining with hydrogen, oxygen, and nitrogen. Carbon fibers are strands of in-line carbon atoms with the highest strength to weight ratio of any material, much stronger and lighter than fiberglass but also more costly. Sheets woven from carbon fiber (reinforcing plastics) are used in racing cars and other high cost applications, such as in aerospace. It combines the strength of steel (or higher) with very low weight. This is most likely the direction of future material developments.

Diamond has a face centered cubic crystal structure of carbon atoms, and is formed out of ordinary carbon under extreme pressure and heat. It is produced in nature in volcanic processes that can now be duplicated artificially. A new but very difficult process is vapour deposition, using methane gas. Diamond has the highest hardness and thermal conductivity of any material. Despite its cost, it is already used in many applications requiring cutting or polishing, and there are many others being developed; diamond is a material for the future. It has amazing heat conduction and can withstand high temperatures, making it very useful in conducting high electrical currents (when doped with boron) to replace silicone which is useless above the temperature of boiling water. An interesting feature is that light transmitted through diamond slows down by as much as 40 %.

ENERGY AND THE ENVIRONMENT
Unlimited and inexhaustible energy will represent a turning point in civilization

Conservation of Natural Resources
Conserve the environment today or it will be
done for us

Our accelerating need for natural resources, such as agricultural land, minerals, forests, oceans, fresh water, and clean air, is outstripping what the Earth can support, and the impact is showing in many ominous ways. It is hardly necessary to mention this here but it would be unrealistic not to do so, the way it sneaked up on us despite early predictions by many experts. If we don't fully heed it now it will impact society severely in following decades, no question about it! Air pollution from burning fossil fuels and emission of greenhouse gases affects people's health and it is the likely reason for "global warming". It is possible that Earth is also entering a natural phase of warming at the same time, and humanity's contribution is then superimposed. It is nearly certain that the climate in the entire world will become unstable, resulting in centuries of severe weather and economic, population, and social pressures. Furthermore, flooding of all low lying areas may change the face of the Earth before natural negative feedback or human intervention stabilizes it.

The need to conserve our environment and wildlife resources will impact our way of life in many ways, ranging from the disposal of waste to clear-cutting forests, pesticides, over-fishing, pollution of rivers and oceans, etc. Careless plunder of the environment is only a minor hiccup as far as Earth is concerned, it will eventually recover one way or the other; but we may not be here, at least not in what we know as a human society. A good example of our international carelessness and disorganisation is the ridiculous practice of whale hunting by some nations, producing excessive pressure on Blue and Right whales that will result in their extinction. Whales are highly intelligent and the hunting methods are (by necessity) very cruel; it takes hours for some whales to die. Weekend protesters are too busy screaming about the killing of cuddly cute animals, dead in seconds and not endangered by extinction. The next time they eat pork chops or steak they should pause and think, but such animals are of course not really cute. Like it or not, our predecessors were omnivorous, and changing that to herbivorous in a hurry is unrealistic. Tasty substitutes may be developed in the near future and maybe no one will have cravings for steak anymore.

Fossil fuels and Nuclear Energy
Fossil fuels must be conserved for future
chemical usage

A replacement for fossil fuels must be found before 2040 to avoid widespread world poverty; oil and gas reserves will then have reached a point where it will no longer support even our most critical energy requirements. Remaining fossil fuels should then be used only for essential chemical applications and become very expensive. Solar and wind energy will help but they are also costly and not practical on a very large scale. Hydroelectric power is efficient but it has limited availability and environmentalists take a dim view of replacing beautiful valleys with lakes that crowd out wildlife and agriculture. Development of mobile electric-power sources that do not use fossil fuels is urgently required for such moving platforms as ships and aircraft. The only immediate solution is nuclear power, but we know from 60 years of experience that fission of uranium is impractical. It suffers from unacceptable radioactive waste, a potential of explosive accidental leaks, and high initial and maintenance costs.

It was recognized 50 years ago that nuclear fusion had many potential advantages over fission, but its practical application has proven elusive. Fusion power releases some of the 'Big Bang' energy, imparted to three of hydrogen's isotopes found in nature (protium, deuterium, and tritium), with zero, one, and two neutrons in the nucleus, resp. Tritium is by far the easiest hydrogen isotope to fuse with other light nuclei (its nuclear strong force is greatest), but it must be artificially created since a short half-life makes its presence in nature rare.

When extreme temperatures (particle velocity) and pressure fuse hydrogen isotopes into a (helium) nucleus, their combined mass (energy) is less than it was before. This energy difference is recovered as radiation and converted to heat, used to generate electric power with steam turbines, etc. The same happens when fusing heavier nuclei up to iron, but with progressively lower efficiency. Fusion beyond iron requires more energy than what can be recovered because those nuclei have more total mass than prior to fusing. This explains why fission (splitting) of heavy atom-nuclei (e.g. uranium) also releases enormous energy. Fusion is the reverse process of fission. Fusing into iron yields no energy at all and neither will fission; therefore an iron nucleus is very stable and not likely to ever change.

Future Energy from Fusion
Energy from nuclear fusion is essential for future progress

The underlying process in any nuclear reaction is that two interacting protons experience a tug of war between their (long range) repelling positive charges and the attracting 'strong force' that acts only over a very short distance. It takes a lot of kinetic energy (particle velocity or temperature) and density (compression) to overcome the electric repulsion between two nuclei with the same charge and get them close enough for the strong force to fuse them. Nucleons (protons or neutrons) are composite particles, each consisting of three quarks dancing around each other at relativistic velocities in a stable relationship. The quarks in a fused nucleus (two protons plus neutron companions) are quanta-waves in stable but wildly variable acceleration under the influence of the strong force and electrical repulsion.

There have been many research projects over the last 50 years attempting to demonstrate profitable methods for achieving net fusion power (more energy out than in). Although there were high hopes for the latest scheme called 'ITER' (International Thermonuclear Experimental Reactor) and its projected follow-on 'DEMO', the break-even point has not yet been reached on a sustained basis. Many difficulties (requiring $ billions) must be solved and a successful conclusion is debatable at this time. It uses the 'Tokamak' design, like its predecessors, fusing deuterium and tritium nuclei in an evacuated toroidal chamber. The gas is heated into a rotating plasma (stripping away electrons) and is magnetically confined (compressed and kept away from the fragile walls) at relatively low pressure or density but at very high temperatures to fuse the two positively charged nuclei.

Another approach uses 'inertial confinement', where pellets filled with hydrogen isotopes (deuterium and tritium) are blasted by very high power lasers to evaporate it fast enough that the gas inside implodes and heats to fusion temperatures. Obviously, this is not a continuous process and new pellets must be supplied automatically for every cycle. Recent improvements have made this again an attractive approach, although it suffers from the same drawbacks as magnetic confinement, namely efficient removal of excess energy and leaking of radioactive tritium. Both methods require nuclear breeding of tritium from lithium, a very expensive and hazardous proposition.

It is hoped that one or both of the above methods will eventually be successful in some form, since it is essential for society and civilization's survival. However, in the long-run something else will be needed and that may be direct fusion of deuterium (proton + neutron) into a helium nucleus (two protons + two neutrons). The temperature required to fuse nuclei depends on the overall electric charge. Fusing deuterium into helium is therefore attractive; also, deuterium is available at little cost. The drawback is that confinement has to be much better and the energy output is much less compared to fusion of deuterium and tritium. Elimination of tritium is the goal, but problems with it still outweigh any advantages so far.

Long-term Nuclear Challenges
The Sun's fusion process may be copied in the
distant future

A final and optimum solution is fusion of protons into helium nuclei in one single compound step, like in the Sun. But this requires extreme confinement or proton beam targeting to accuracies not even dreamed of today. Charge repulsion during approach is the main roadblock, two protons always missing each other, except when aimed dead-on. In today's processes only a very small fraction will approach close enough for the nuclear strong force to take control. It is the same problem in balancing a sharp pencil to stand on its point; in theory it can be done, at least for a short time when all disturbances (including thermal) are eliminated, but not in practice! Future science and engineering must concentrate on developing separated beams of protons and electrons, focused and collided at unheard of accuracy (like balancing a pencil on its point!). Liberated energy from fused helium in the form of radiation is then used to generate electricity. Development of small accelerators or plasma toroidal-rings with extreme focusing accuracy and power is the primary engineering challenge for our descendants, but probably not realized until the fourth millennium .

It is my guess that Scientists and Engineers will eventually solve these problems and that Humanity shall never look back from this early period in history when the use of free, unlimited and inexhaustible portable energy was only a dream. It will represent a turning point for our civilization and open doors to a future we cannot yet imagine. It should free us from our Earthly confinement and initiate large scale human galactic exploration.

TECHNOLOGY AND LIFESTYLE
Science and Technology are tools for the achievement of Society's objectives.

Data access and Development
New technologies open the door for
intervention in daily life

Information Technology is the great technical upheaval of the second half of the 20th century, continuing into this century without let-up. The collection, storage, processing, transmission, and communication of data has revolutionized our lives and it is widespread enough that elaboration is unnecessary. That suits me; I worked for a company engaged in electronic development and managed some subcontracted software development I but could not comment intelligently on today's products. Most people have a rough idea of what it does, how it does it, and where it may lead, but the complexity makes it comprehensible for experts only. The rate of development has been exponential and that is expected to continue. This is worrisome because subconsciously we always need to know where information comes from, otherwise it is not trusted. The combination of instant communications and information overload have made workaholics and zombies out of many. On the other hand, most people learn to protect themselves and selectively use these new technologies to make life easier. And that doesn't mean the addicts calling their spouses (or partners) in a supermarket to discuss what is on special that day.

To what extent information technology will change people's culture and behavior in the long term is questionable. Family and social life changed already and the door is now open for direct intervention in daily activities. You may be employed by a centrally controlled organization where all duties are regulated. Your detailed instructions, scheduled activities, and communication may be transmitted to you and recorded. You may be fitted with a device that informs authorized people where you are, what you are doing, who you meet with, etc. There will be little privacy, except going to the washroom probably, and no escape unless you opt out and suffer the consequences. You think that is exaggerated? And if you don't like it someone else will do it for the sake of a secure existence. It is a good technique for keeping track of habitual offenders; simply put them back in jail when they go off the air. You can withdraw and ignore it, but how do you compete? You'd lose contact with the rest of the world, and consequently the future for individualists, is not very bright. We should be concerned about what it will do to people living several centuries from now and beyond. This sort of thing should be outlawed, for humanity's sake!

It is pretty obvious where things are going; you only need to look at what happened to cars in the last 20 years! We'll be inundated with automatic wireless sensor-chips, collecting data and continuously comparing performance against predictions, and project trends. You won't drink water unless it was chemically analyzed, and you'll know at all times exactly where your family-members are. Disgusting!

Benefits and Obligations
I.T. systems can now control and prevent any
social excess

There are of course many benefits from instant communication and access to data, it allows timely decisions. Human emotion and motivation won't change much and the challenge for future leaders will be to control it all when things will be far more complicated than they are today. Observing society for the last fifty years it seems that the direction of global development is probably the best we can hope for, it could have been much worse. The catastrophes we witnessed (on television), mostly genocide, natural and economic disasters, famine, floods, etc, were aggravated by social inadequacies at many levels and were often preventable, although not without hindsight and a collective will to make difficult decisions. There are no obvious political routes towards balanced social policies, just like hockey games cannot be planned in any detail beforehand. Even with a crystal ball, the obstacles of vested interests and prejudices are nearly insurmountable. Therefore such interests should never become too big!

It is much easier to identify what will not work. Communism lacks personal incentives, and unrestricted capitalism leads to escalating separation between rich and poor, like in previous centuries. That leaves democratic social systems, perceived to support slackers at the expense of thrifty people. But it doesn't have to be like that; today's Information Technology (I.T.) allows a functional blending of the best capitalistic and socialistic principles. What may work is a fair and foolproof taxation system, with no easy loopholes.

Present tax-systems were developed by changing stop-gap measures every taxation year and it is high time politicians find the courage to start from scratch. It should be uneconomical tax-wise to pay top executives a thousand times more than an average worker, because that revives elitist or aristocratic systems in everything but name. Bartering between individuals requires no controls; but employment paid for with untaxed goods or perks (and charged as a business expense) should receive enforced penalties stiff enough to stop it entirely. Essential social and medical provisions should be the same for all; administered and paid for by governments out of taxation.

The desire to live a life more comfortable than what minimum social benefits provides is strong enough to motivate everyone except the incorrigible. Only confirmed handicapped people deserve special assistance in such a system. Many have the opinion that if you are too lazy, addicted, or otherwise incapable to work for a living, you deserve to starve. Yet, when a son or daughter ends up in such a predicament, most families support them. Do people not receiving such support deserve similar fair treatment and compassion? Any sub-society with abandoned people starving or freezing to death is pitiless, and their citizens lack compassion or are reluctant to speak out! Not everyone is blessed with the mental and physical resources to live unaided in this complex world and we have an obligation; but the rules should be clear to all!

Colonization of Outer Space
Defilement of Earth will force space colonization in the future

Assuming that no economic catastrophes occur, exploration of space will proceed slowly. The reasons may be energy, resources, national security, scientific, curiosity, or possibly survival. It could even develop into a viable tourist industry. That is all very interesting but it hardly requires our attention; what will be, will be! However, space exploration has a role in the philosophical projection of humanity's future. Humans will never be immortal or extend their life span enough to explore distant space and return to Earth, although we don't know how deep-hibernation methods will develop. There are several options: -- perpetually wandering human space-colonies, with their own isolated civilizations, or -- exploration by sophisticated artificial intelligence designed by humans and reporting back if and when feasible, or -- anything in between.

Projected over the very long-term there are really no obstacles against any of it. Not that it is bound to happen because we may become extinct before then. God never did like how the age of reptiles turned out and bombarded them with a huge meteor 65 million years ago! Anyway, nature will eventually provide us with unlimited energy in the form of nuclear fusion, because hydrogen is everywhere in space. Apart from (recyclable) raw materials, relatively abundant on asteroids, energy from fusion is the only essential prerequisite for unlimited space travel. We will eventually learn to produce and store anti-matter more economically than today (at a billion dollar per milligram), but it is the most concentrated form of energy. It is not really needed because unlimited fusion-energy can accelerate both hydrogen and helium close to lightspeed, expelling it at many times its normal mass.

What is the objective of such space exploration? This is philosophical because it matters very little what we do, our direct descendants on Earth will never see much benefit from it. It clearly goes against our grain to do something that has no benefit in the near future. So why do it? Because we also have an overpowering instinct to survive, reproduce, and advance, and that should convince us it is nature's plan. It seems obvious that limiting our species to this planet will result in extinction well before ~100,000 years, one way or the other, and that is a compelling reason to colonize space, since there is no alternative. Unless we don't bother to think about it! My guess is that evolution will eventually make it compelling, like our need for love, sex, and friendship, because evolution somehow provided all the other instincts that caused our line of ancestors to survive.

However, it is time for humanity to wake up and stop expecting nature to hold them by the hand. As individuals we are sophisticated enough to set our own goals. The question is if our fragile global institutions are able to overcome the leg-chains and weights of nationalistic and mega-corporations vested interests! This won't happen overnight but it requires people everywhere to think and talk about it, especially in the media. To become aware of the inertia and dangers, to stop admiring patriotic hoopla and develop pride in human achievement. It can be done, but people don't yet realize that it must be done!

Earth shall remain home base, but it may only be familiar to the vast majority through historical records. It probably involves a functional mixture of human and artificial intelligence, looking ahead 5 to 50 thousand years from now. One exciting (but small) possibility is an encounter with other terrestrial civilizations, and then we may also be recorded in galactic history. Of course, that matters about as much as it did to Charlemagne to be mentioned in today's history books. And as far as artificial life is concerned, do you know anyone who really likes artificial flowers, never mind how perfect they are? What fascinates us is the mystery of nature, and that we are here to observe it, and part of it!

Assuming we survive for the next 500 years, it is probable that the obstacles of one-way space travel will be worked out. We've done it before, in the 17th and 18th centuries, when 99.9 percent of the people who emigrated to the Americas, Australia, or elsewhere, never returned. This is why I like science fiction; imagine the scenario of thousands of people boarding spacecraft to transfer them to huge orbiting communities, slowly accelerating towards star systems centuries away. They will die on it, with the knowledge that only distant descendants may some day land on an alien planet as colonists. But stranger things have happened, and no one on Earth will ever know what became of them! You believe it is all far-fetched, and pure fantasy? Of course it is fantasy, but it is also in our genes; if our ancestors had not ventured out of Africa we probably would have been just one of Nature's unfulfilled

142

potentials. Stop and consider how far we have gone in a mere 200 years.

Medical research and Applications
Technology is changing medical procedures and treatments

One field of technology, more useful to society than almost any other, is the development and availability of new medical drugs, diagnostic testing, and treatment procedures. We all know family, friends, or other acquaintances who have been tested or treated for illnesses using MRI, CT-scan, chemotherapy, angioplasty, and other medical procedures with strange sounding names. Modern doctors still diagnose their patients primarily on symptoms, but they now have the option to request a growing number of procedures to confirm their diagnosis and decide on either traditional or new treatments in consultation with other specialists. This has changed healthcare completely, but it also created a major social problem because of their complexity and cost. This is due to expensive drugs, specialized medical staff, and exploding costs for administration, hospitals, and equipment. The loss of productivity and cost of care and treatment for millions of patients today has become a major burden on society although compassionate considerations are still a first priority, at least in most places.

One new development worth singling out is the controversial use of embryonic stem cells for research, basic human cells not yet triggered by the body to divide and specialize. Ideally, after receiving the necessary chemical signals from living tissue that stem cells are injected into, it makes unique protein to rejuvenate any organ or tissue in the body. All embryos initially consist only of basic stem cells, whereas oriented (adult) stem cells are responsible for the growth and subsequent renewal of specific functions of a child's or adult's body. This subject is controversial because pro-life proponents argue that the use of stem cells from human embryos destroys potential human life.

There has already been considerable success with adult stem cells but US government funding for the use of embryos has only recently been approved. Stem cell research is being carried out to varying degrees in many countries. Embryonic cells are potentially more versatile (especially for conditions related to the nervous system) but also less tolerated by the human immune system. It is considered promising for the treatment of serious diseases and injuries such as

143

Juvenile diabetes, Cancer, Cardiac damage, Parkinson's, Muscular dystrophy, Multiple Sclerosis, Burns, Strokes, and Spinal cord damage. It may unfortunately be a long time before such benefits are realized.

Health care and cost
Social responsibility includes taxation for
uniform health care

The best example of the usefulness and impact on society of Science is how it affects health care. Advances in electronic sensors, digital computers, and display devices, combined with the exponential growth in drugs and medical knowledge, make lay-people wonder how doctors keep up with it all. The general public is reasonably informed because there is much curiosity and frequent coverage on television. The rapidly expanding science of genetics produced a lot of new (and expensive) chemical drugs, with many older ones discouraged due to recently discovered side effects. This frantic activity attracted many venture investors and increased medical costs for patients with serious problems, sometimes to a breaking point.

Various semi-socialistic governments made brave attempts in the late 20th century to help the poor and elderly but are now looking for ways to get out of a crippling financial trap. There is no doubt about the benefits of modern medicine, but the financial impact of serious illness is has become an almost irresolvable problem. The option of reduced treatment is repulsive, but the alternative today is rich investors profiting from the financial ruin of poor people or governments. Historically such dilemmas caused political upheavals and it is true, health policy is clearly a government responsibility. Medical care everywhere urgently requires social legislation to eliminate complex administration and financial approval cycles. The only realistic solution is to provide basic and equal treatment for all with rigid guidelines, paid for by fair taxation. Anyone wanting more can do so at their own cost.

The battle now going on in the U.S. over health care, is a fundamental dispute between capitalism and socialism. Both have good and bad features. One side claims health care is a commodity and only free enterprise provides it efficiently. The other side claims that free enterprise will maximize their profits with ridiculous fees that impact only those who can least afford it. Like everything else in nature, the answer is clearly to take the middle ground, because both sides are right to some extent.

Doctors are part of the problem because they consider themselves independent business people, charging whatever the market will bear whenever they can do so. Since professional associations control medical qualifications they also control supply, and this is not unique to doctors. A capitalistic approach is useful because market demand ensures availability. But health care should never be a monopoly; there is room for doctors with paid for education to practice as government employees (with their own union of course). Obviously such a scheme entails paying it all back when quitting government employment. The government has a social obligation to provide affordable basic health care for everyone, and a mixed system is perfectly acceptable, with proper controls. The profit from educational effort and investment does not entirely belong to an individual either; a part belongs to a Society at large that established the means and provided facilities for obtaining such an education!

It is a peculiar mixture of compassion, social need, guilt, and greed that drives modern health-care. People now live longer due to greater affluence and medical-care, but many develop serious and costly problems from indulgent lifestyles, bad diet, and a lack of exercise. Visiting a doctor for every problem is always encouraged and care of ailing elderly often becomes an agonizing nightmare of diagnosis and treatment. In previous centuries old people accepted their fate that aging and its associated problems is unavoidable. They prayed for a death with dignity and without suffering. Our system today is based on compassion for the patient and for the patient's family who often won't accept the reality of impending death. They understandably insist on maximum treatment, but that frequently prolongs unnecessary suffering. Doctors support it because they are humane, and with life there is hope (and they get paid for it). It seems heartless for mere statistics to decide if treatment should be stopped to reduce the period of suffering. An affordable environment that is more comfortable for the patient should then be available, although it is often impossible to convince elderly people they are no longer capable of living independently.

Lifestyle and Materialism
Technology has been good for the economy, but it changed our lives

This is probably a good time and place to comment on changing lifestyles, caused by our advanced industrial technology (as if you don't already hear enough about that subject!). How and why has our lifestyle changed so much in the last 50 years? First, let me mention that 50 or 60 years ago average working people in the western world spent their income primarily on food, house-rent, and clothes. A smaller amount usually went for things like camping in (one week) summer holidays, fixing or replacing your bicycle, education, and entertainment like going to the movies once a month. Today the biggest slice by far is the house mortgage or rent, then food, followed by transportation, entertainment, health care, children's education, and clothes, etc.

To elaborate: a family of four used to rent a townhouse of less than 1000 square feet. Many families now dream of owning a detached house of 2500 square feet, with a double or triple garage to park their SUV's, boat, and snowmobile, in addition to thousands of dollars worth of children and adult toys and sports equipment. No household is complete without personal computers, several large-screen televisions and all the gadgets that come with it. It is not a holiday unless you spend thousands on skiing or soaking up the sun in the winter, on a cruise, or at your own lake cottage in the summer. Is this just the opinion of an old man, spiteful that my youth did not provide such a lifestyle? No, not really, I like these things as well, given the opportunity. The question is if such expectations are sustainable in the future, and if it remains a lifestyle available only to fortunate people in even more fortunate parts of the world? Don't get me wrong, this is not a condemnation of new technology; it has been beneficial, and widespread enthusiasm for cell phones, computers, etc. has made their development commercially viable. It has been good for the economy and for technological progress, but has it been good for us?

My recollection of adults 50 or 60 years ago is that they were more content; not as restless as people today. An abundance of opportunities and choices now entice you to take advantage by copying friends and neighbors. Fifty years ago a job was often more meaningful, although probably less challenging. It was not an every day nuisance to make money, requiring a car and a one-hour drive to get there. Most people came home for lunch to join wives who did not work (not passing judgment here, just telling you how it was!). People now shudder at the

146

thought of such a simple lifestyle but if you don't expect more, you don't miss it. Sure, there were rich people then also showing off their money; and the really rich had a lifestyle that separated them completely from the masses. They probably felt out of place at seaside beach holidays (without any doubt my best memories of all) because they were always hiding in luxury hotels! There are options; the future does not need to be more of everything with new technologies or concepts providing ever more choices intended to dilute stress in our lives. The trend today is to manufacture everything in the world's cheapest labor markets, and that is probably good. Hopefully spreading of the wealth continues, elevating such regions out of poverty and improving education. And that is what is needed!

CHAPTER V - <u>APPEARANCES OR REALITY?</u>

IS THE UNIVERSE A HOLOGRAM?
We live in a connected Universe, but how we are connected we don't know!

<u>A Connected Universe</u>
When you go for a walk the entire Universe goes
with you

This section, about holograms, won't make much sense unless you have some familiarity with the esoteric concepts discussed previously in "A Vacuum Analogy". If not, you should skip to the next major section.

In our frame of reference we consider ourselves king; it is a perfect world for us to observe, using photons as messengers. But wait a minute, what we call particles are only collecting and redistribution points for these uncountable intermediaries or energy quanta-waves that zip around in the Universe. Is this Universe some sort of dynamic memory for something else, existing at another level or in other dimensions? You hear about that sort of thing almost every night somewhere on TV, and most people never look at it, so why suggest it here? But consider this for a moment: without photons our world would be static and basically dead; it actually could not exist at all. Particles are self-perpetuating fluctuations in the Vacuum, but without long-range photons there is only gravity and it would crunch everything together in a big hurry.

Photons, with energies ranging from huge to miniscule, are responsible for what we sense. As far as we are concerned the Vacuum does not exist because we only sense discrete energy-fluctuations called particles and photons; they give everything form and identity. On a slippery slope already, it would be risky to suggest that our Universe(s) is a memory for something we are an integral part of, as intelligent observers. Evidence for every event that ever happened in our Universe is still present in one form or another, resembling a memory. That is going much too far of course, but it is interesting just the same! One idea I do want to stress is that clearly everything (particles, photons, air, water, fire, etc.) is connected, never mind how you look at it, otherwise there'd be no forces, gravity, electricity, light, sight or sound. I don't mean to give it supernatural connotations but we are connected beings, although unaware of the connections! And when you go for a walk by yourself, the entire Universe goes with you! OK, laugh, but it is true!

Holograms and Instinct
Instinct connects us to otherwise inaccessible information

Physical matter (biologically alive or inert) is analogous to a physicist's hologram where interference between two split laser-beams, reflecting of an object, is recorded on a photographic plate. When these laser-beams are directed through this plate they can project a three-dimensional image of the object on a screen. What is important, and analogous, is that any small area on this (two-dimensional) plate can recreate the entire (three-dimensional) image. An ordinary picture is a composite of many small separate pictures, but a hologram holds the entire picture in every small region, although resolution increases with size. The analogy with physical matter is that photons created in a single event remain instantly connected (without any time lag) and therefore represent what may be considered a historical image of the event.

Expanding on that, energy-aberrations distributed in the Universe represent an instant record of all events that ever took place in the entire history of the Universe. However, we can only tune in to what originated in our ref. frame, like Alan Aspect showed in Paris in a breakthrough experiment in 1982. Two photons (from a single event) remain connected as far as we are concerned and together as far as they are concerned. It shows that our concept of space is real only in our own ref. frame. Einstein would have been amazed, but it resolved his argument with Niels Bohr. He simply did not extend his relativistic concepts far enough!

Since photons (real and virtual) determine the movement and location of particles outside of nucleii, the composition of any macro structure of matter (such as a human brain) in principle stores within any part of its volume some representation of everything around it, with longer distance and smaller size reducing the resolution. And it is therefore not difficult to accept that the Universe is in very rough terms (and only in principle) a huge memory of all that ever happened within it! Any photon originating on Earth in a barbeque and traveling through the Universe at lightspeed near some other galaxy, one hundred million light-years away, will reckon it is still located inside the barbeque; but only in its own ref. frame! We are connected, but at a

149

level inaccessible to our consciousness and therefore we have the conviction that we are completely separate entities as individuals. Of course, we are separate, but not entirely! It may be that ants (with very little consciousness and no logical reasoning) have some access, allowing an ant colony to act with single purpose and direction.

It may also be that our senses, logical reasoning, and consciousness lock us into our version of reality, and our frame of reference. It is possible that instinct provides unsuspected access, since some demonstrated abilities are a deep mystery otherwise (e.g. Monarch butterflies). It is irrational to suggest that a connected Universe could be responsible for the formation of life, but copying of complex molecules (e.g. proteins) may simply be more probable when others are nearby to be copied. The information necessary to replicate biological structures is present (in principle only) in any volume of the hologram-like Vacuum, and maybe only the raw materials (e.g. amino acid molecules) and a control system such as DNA/RNA is required.

Bohm and a Holographic Universe
Scientists have found evidence of realism in
metaphysical concepts

David Bohm, in addition to 'The Undivided Universe' (1993, Routledge, written with B.J. Hiley) also wrote a speculative book called: 'Wholeness and the Implicate Order' (Ark, 1980), in which the concept of the Universe as a hologram is discussed philosophically. To show that this is not as preposterous as it appears, the following quotes in abbreviated format outline some of the comments made publicly by Bohm and others on his unique philosophies:

- *Bohm has said that our senses are lenses, and if you take the lenses away you've got a hologram. Lenses tend to reify, to objectify and articulate particles. Take the lenses away and you've got this distributed.*
- *Despite its apparent solidity the universe is at heart a phantasm, a gigantic and splendidly detailed hologram.*
- *In his book "Holographic Universe" Michael Talbot suggests that a paradigm shift in our science and culture is at hand.*
- *What we think we know is often incorrect or incomplete.*
- *Mathematics is not immune from contradictions in logic when forced to accept probabilities.*
- *Putting the holographic structure of reality together with its perpetual dynamism, we get the holomovement: an exceedingly rich and intricate flow*

in which, in some sense, every portion of the flow contains the entire flow. As Bohm puts it, the holomovement refers to 'the unbroken wholeness of the totality of existence as an undivided flowing movement without borders'. The physical evidence that forms the basis for postulating the holomovement comes primarily from Bohm's interpretation of physics, especially quantum theory.

- *Fundamentally, the particle is only an abstraction that is manifest to our senses. 'What is', is always a totality of ensembles, all present together, in an orderly series of stages of enfoldment and unfoldment, which intermingle and inter-penetrate each other in principle throughout the whole of space (Bohm, 1980).*

- *Bohm's understanding of physical reality turns the commonplace notion of 'empty space' completely on its head. For Bohm, space is not some giant vacuum through which matter moves; space is every bit as real as the matter that moves through it. Space and matter are intimately interconnected* (a concept also endorsed by Einstein).

Much of the foregoing is classified as metaphysics, the part of philosophy that investigates ultimate reality. It was historically known as 'natural philosophy' and limitless unless restricted by defined boundaries. Although there is little doubt that everything is connected at some low level which cannot be sensed or comprehended, any supernatural connotations are entirely speculative and vacant of evidence.

WHAT IS IMPORTANT, AND DO WE CARE?
Emotions, assumptions, and vested interests are the cause of all discord in society.

<u>Happiness and Instinct</u>
Social relations are more important than
conquest or wealth

It is a bit foolish, but if we know and care about what is really important to us humans are unique; all other creatures on Earth may care but they don't know, instinct looks after that. Yes, we do know and care (most of the time!) because we want to be happy and healthy and most of us would also like to be wealthy. We conduct our lives accordingly and make compromises to achieve it. Individual happiness in any society involves making others happy also and we should thank evolution for arranging that, because it is crucial. Almost all our desires are based on instinct or pre-programming, and that applies even more for other high orders of life on Earth. Adults in their prime have a social responsibility to recognize the feelings of people close to them, especially the young and the old. Although children and parents are generally revered, they mostly find any emotional companionship outside of the family. This is normal, but it can cause unnecessary emptiness. Grandma may not be interested in talking about the latest fads but she can tell you why you are not shaking off that cold, in addition to many other things where logic is blurred. And children flourish when you give them a chance to state their opinion and add to family conversation!

One theme of this book is relationships, socially and in the world around us, and why we take so much for granted as obvious and real. However, we have seen that reality is nebulous; what we sense is real enough, but it is always our interpretation of a projection on our mind. We don't worry about that, nor should we. Attempts at discussing reality are made in a roundabout way by the news media and at scientific, religious, or social gatherings, or during a crisis period when we unsuccessfully try to make sense out of sad and unforeseen events. Then we reflect from a remote vantagepoint, to protect ourselves. Psychologists studied the human psyche and developed many theories to predict what we are apt to do in specific circumstances. We don't really care much, finding it interesting or good for a joke, but it is never a factor in decision-making. When something is really important we turn to intuition or emotion and apply logic later, just to make sure that what we are planning doesn't blow up in our faces. In any case, we are more likely to override logic than go against any deep emotional preferences.

One relationship we talk about all the time is that between genders. The role of women has changed completely in the last 100 years or so. Not one today, at least in western countries, looks upon her husband or partner as her lord and master. In fact, men may still be dominant in business or in specific vocations but women mostly control the daily social life. And if a man doesn't fit in he is left to watch sports on the TV, with others like him. Women have a better grasp of society and cooperation than men do; consequently men may need to adapt more in the future. Only those with a useful role to play can have real status; maleness by itself no longer guarantees it.

Society, Justice and Compassion
Society must learn to compromise without hurting people

Intuitive decision-making, geared to relationships, was well suited to humanity's climb from cave dwellings to pre-historic village life. It was not very helpful in the period of kings and religious rivalries when individual freedom and happiness was often minimal. And it is not ideal now, in our complex and huge modern society. Families don't need to rally around one member anymore to protect him or her from irate neighbors, that is now the job of the police. Mobs or arbitrary rulers are no longer in charge, with laws protecting us. Hysterical lynching, beating, or stoning has been curbed in most places, although victimized people everywhere still crave for such revenge sometimes. We will never be like robots in a beehive society; what makes us unique is our logical and emotional inclination of demanding equal deference to justice (deterrent) and compassion (mercy). How can such fairness be accommodated in law? It often provokes the disorder we see in the newspapers every day, when protesters battle with police for instance. Even super-logical artificial intelligence (A.I.) could and would never create such a society; only humans can with instinctive motivation and fears, balanced by logical and moral impetus.

Given enough time A.I., might evolve a society of robots, pre-programmed with predetermined objectives, like ants, and never wondering about any purpose. Even imagining such a thing is depressing! All in all, and considering the obstacles, our ability to compromise between law and human rights is miraculous, but it is also the reason for unavoidable human conflict. The most basic law of nature is duality, stabilized by two separate coercions that define each other. The important question is if we can learn to compromise without hurting people! What has been achieved should convince us not to underestimate what world peace, cooperation, education, and economic unity can achieve. Society must protect and move everyone forward, and not just the rich. Human society should not be used to advance unfair individualistic gain, it must serve all people equally. And we must reject partitioning, encourage enlightenment, and overcome the obstruction of civilized society by ignorance.

Human rights and Vested interests
Civil rights and the environment should be
prime objectives

All this is easy to write about, but grand suggestions for social improvement are always impossible to implement. They have trivial effect because power to change resides with those upholding the status quo, the base for their power! Practical solutions must evolve slowly, from the bottom up. Social improvement in many places is clearly urgent, but our highest priority must be to safeguard what we already have! Civilizations were never robust and we cannot assume that progress or even stability is typical. Like everything else in nature, affluence is invariably a cycle and can easily drop so quickly that reversal is impossible (even during an upward swing). A major nuclear war may be more than a temporary disaster; it can trigger a slide towards extinction of Homo Sapiens! Prosperity and tolerance encourages negative traits in human nature, when selfishness seems more relevant than compassion and invites division and dogmatism. It is typical for any society not preoccupied with survival to destroy itself through intolerant partitioning. Every civilization in history managed to do that, although most were not really civilized by today's standards. Our obvious task is to reverse the trend and establish a true global society!

Therefore, protecting and expanding what we have is most important, and evolution (biological and social) will take care of the rest, with our help of course! What ought to be the ground rules? That's easy: civil rights and protection of the environment are paramount, demanding more than just lip service. Abuses today are widespread, even in supposedly liberal countries, and authorities get away with exercising selective powers by using conflicting and confusing laws, and by purposely failing to act decisively. This is generally not corruption but a fear of destabilizing society; exactly what is needed at times. Media attention is effective and essential to counteract some often callous violations. Human rights and fairness should have equal billing with criminal law, but unfortunately judges are often inconsistent in applying the letter of the law. Circumstances should always influence leniency or vengeance, although it is the interpretation of such circumstances that is the problem.

Religious influence over justice can be erratic, the concept of mercy is applicable only to those showing some religious empathy. However, citizens must know what they cannot do under the law and be relatively certain of the consequences. Children must be indoctrinated with a strong sense of ethical behavior; any misbehavior must never be ignored for fear of hurting their feelings. Comfortable people must participate in society and reject a "I have what I want, and good luck to you" attitude! Excuses for suppression of civil rights may be 'the country is not ready for it', or 'it will result in chaos', or 'the economy cannot afford it'. There is usually some truth in it, but it provides leaders with easy excuses for the status quo. Liberal leaders, and more flexible laws, will never happen unless forced. And as far as environmental guardianship is concerned the evidence is obvious, requiring not much elaboration!

Successes and Failures
Life is a succession of satisfying objectives and
desires

Having reached an age where ambition is waning, it is interesting to assess the successes or failures of decisions made in life, when there was a choice. Personal choices are most important and it often depends on opportunity and luck. Happiness and good health also depend on that, but you can improve the odds as a young person by looking around and noting mistakes other people made. Deliberate choices are difficult to separate from forced or random decisions. We combine emotional desires, personal interests, and capabilities with ambition and opportunity to achieve goals of happiness, money, prestige, and status. This generally results in some success, although nothing may stand out when you retire.

You remember the satisfaction of success, because it was the prime motivator. Secondly, you remember people you were friendly with, although many you'll never see again. Such are the memories you take with you, but that is not enough; you must have other absorbing interests (like writing a book!). Young people do not quite understand how important it is to select a fulfilling occupation. You can be a bookkeeper or organize rock-concerts and be completely satisfied years later, but only if it suited your personality and capability. You will remember the best and the worst, and inevitably there will be a lot of the latter. Such memories make you happy, sad, or embarrassed for the rest of your life! At a young age it is difficult to imagine the world without you, but it becomes an issue later.

<u>Discord in society</u>
Opinions and greed caused all of the major wars
in history

Is there a way to short-circuit evolution and make human foibles less destructive in modern society? Religions usually prevail on people to override their inborn selfishness and craving for socially harmful excitement. This was successful at many levels, although religious dogmatism was also responsible for some of the most repulsive and bloodiest wars in history. Given humanity's recently acquired proficiency to annihilate anyone and anything, religious fanaticism is more dangerous then ever, although the same applies to racial and territorial exclusion. As mentioned before, evolution works best within many small isolated groups. In fact, biological evolution may not be much of a factor in today's global society; it is simply too big and chaotic. Wishing for another intelligence or consciousness boost, as occurred a few times over the last two million years, may not be realistic or even desirable. Neither should we copy Hitler's attempt to create a genetic 'super-race'. It would hurt far more than help, aside from the fact that it is absurd, immoral, and revolting. All we can do is keep our fingers crossed!

If we need an example of shameful collective human behavior, the start of the first world-war in 1914 does very nicely. The trigger-events were simply excuses, only tipping the balance of hundreds of secret deals and understandings between European nations. Germany in the late 19th century had grown into a major power, economically and culturally, after its many minor realms had merged under Prussian royalty. Unfortunately, the old powers (England, France, and Russia) did not allow it to enter their comfortable colonial club. German elite, aristocracy, political authority, and to some extent the general population, pushed each other into an emotional frenzy over this, and other perceived affronts.

The idea of a cleansing war became popular in Germany, to the point that the actual beginning was celebrated as a victory. Although the real instigators were the German army and navy, 'kaiser' (emperor) Wilhelm II gladly accepted figurehead leadership since he had a score to settle with his close family, the royals in England, who shunned him. We know the outcome; 16 million people died in this stupid war; 6.8m. were civilians and the rest initially enthusiastic volunteers who lost their lives in an unbelievable blood bath, for glory and patriotism. Germany lost, left with conditions so desperate that Adolf Hitler decided to take revenge and do it all over again in WWII.

And here is another obvious problem: we typically choose charismatic leaders who are attracted to power and supported by vested interests behind the scenes. A better approach might be for elected political representatives to select a leader from among them after each election, with an annual vote and mandatory replacement every four years. Or any other practical set of rules that limits the influence of vested interests and prevents 'deals' from creeping in on the edges of each election. Why should we emphasize human rights? Because we are not ants, willing to act unquestioningly for the good of the colony. That is an outdated patriotic and emotional holdover from the Victorian era; before that hardly anyone volunteered to die for his or her country, unless they were starving, and certainly not at less than minimal wage. They were either drafted (forcibly), or hired for a good wage as soldiers of fortune!

People must be involved and party to a decision if some crisis is worth dying for. Issuing orders to eliminate objectionable elements within specific groups is sometimes necessary as a defensive move but it carries a huge risk of getting out of hand, resulting in atrocities and killing of innocent people. The point is that citizens must install authority that does not overpower individual rights, friend or foe. If this ties their hands too much then other steps must be taken to achieve it, regardless of the cost! Coming back to my silly title: do we really know or care what is important to us? Well, we do and we don't; it depends on how we judge the importance, logically or emotionally. Many leaders don't care enough to spend much time thinking about it, they are too busy preparing for the next election.

This section will close with more motherhood thoughts about free will and decision making. Each decision, big or small, requires motivations that involve both instinct and logical reasoning. We can satisfy selfish desires or creative impulses, create social advantages or act to make others and thereby ourselves feel good. It is crucial to become skilled at thinking about it all beforehand and avoid impetuous decisions. Unless a judgment is logical or mathematically irrefutable, it is subjective. It is not humanly feasible to comprehend and assess all the uncertainties associated with so many options at the same time. We therefore make assumptions to minimize the uncertainties, and then use our imagination to explain the reasons when asked.

Religion falls into this category; some people accept dogma that satisfies their emotional needs and deliberately ignore questions of logic, while others insist that only logic has any significance. Many broad concepts we rely on are subjective and limited in application. We probably accept them only because of cultural indoctrination and emotional need, or to avoid disapproval. However, such dogmatic assumptions have been the cause of much discord in society. The conclusion drawn from rereading the above incoherent ramble is that people really have the combined capability and resources to create a society worthy of the name; but motivation is the question. Whether or not vested interests are defeated is up to us; it could be a test!

MIND AND CONSCIOUSNESS
Consciousness is like watching a movie, except you can affect the outcome.

Consciousness and Free will
Free will allows choices between many imagined
alternatives

Now we can attempt to answer one of the major questions posed in this book: 'is what we are conscious of real, or is it all just a projection of something?' Of course, you should ask what qualifies me to answer that; certainly not my education which was exclusively technical. A large number of academics (with appropriate Phd's) have published extensively on the subject of consciousness, with opinions varying as dictated by their training, inclination, and imagination. Having read some of it, not much stands out as interesting. This is not going to be very interesting either, so why include it in here? Simply because there isn't a better way to focus then by forcing yourself to commit to something. Also, because this question is fundamental to our mysterious existence and only people going through their lives without time for reflection will ignore it completely.

OK, let's get on with it. Philosophers have debated the issue of consciousness interminably and those alive can be divided into three camps: religious, agnostic or atheist. The ones I sympathize with are the agnostics; they sit on the fence and look both ways. Why so little progress, other than a lot of arguing from every possible angle? In a nutshell: because science has not cracked the nut yet (and may never!).

It is obvious that our senses are part of an unbelievably complex energy exchange system. When you look at or listen to (or smell) your dog, your senses perceive and convey signals to parts of the brain where images of your dog are stored. This image is not in one location but in regions, with smaller regions storing the same entire image as a pattern, except not as clearly. Short-term memory relates what your dog is doing at that moment to this stored image and to all sorts of other things, such as: when was he fed the last time? Anyhow, your dog projects itself on your consciousness, making itself part of your reality at that moment. That it does so by means of photons via your eyes, or by air molecules vibrating in your ears (or a smell in your nose!) is irrelevant, unless you decide to think about it. We can dispense with a lot of irrelevance by accepting that what we experience is always a projection on our mind and that the dog is indeed very real, consisting of the latest version of matter as defined by physicists and biologists (unless somebody has created an amazingly good simulation). It is also located exactly where you think it is, although someone traveling at close to the speed of light wouldn't agree. Anyhow, what's the big deal?

Consciousness was always a problem for scientists, mainly because they didn't understand it. Philosophers are a rare breed these days; they were expected to solve this problem but never made a dent. Anyone can tell you that being conscious means you are not asleep! It is not an important issue because many animals are conscious, may be even more so then we are. We once had a dog with an amazing sense of what went on around her; she was just not as smart as we were. Her free will was rather limited because instinct and emotion completely overpowered her logic. We look ahead by hours, days, years; but she only worried about the short term, although I can recall incidents challenging that. She knew exactly when everyone was coming home and certainly knew where her cookie tin was.

Consciousness is often described as an ability to know your own thoughts and actions. Science is hung up on it because it resides in the brain and that should be just another organ in the body and follow nature's laws. Of course, it isn't just another organ, it is a holographic recorder and random thought generator and processor; eventually we may learn all about it. We also have something called 'free will', allowing us to decide if we want to do A or B. All other organs in the body can only do A and follow nature's laws. What is special about the brain? Well, it allows us to imagine our world with either A or B. With stored sensory recall there is some fake reality in it and we choose what we like best. That is consciousness and free will, you either eat your cookie or you don't. It is like watching a movie, except you can make it real and take part. We may decide there and then that we are going to try and kiss the girl; consciousness sets us free, but only if our emotions and instincts allow it!

Let me say a few words about Philosophy. Not that there is anything wrong with it, a philosophical discussion is probably enjoyable if you have the training and obligatory memory to quote all those significant statements uttered over the millennia. My unenlightened complaint is that, as far as Science is concerned, it seems mostly irrelevant or of minor significance at best. I have yet to pick up any philosophical essay and find any part of it interesting. Of course, that probably says more about me than it does about philosophy. But it seems mostly directed at scoring points with other philosophers. It may be relevant in social sciences or religion, although a systematic and critical review of religion is bound to become dogmatic somehow. Philosophy applicable to physical science is usually limited to the significance of any new theory relative to older ones, or society. Interesting as that may be, it never contributes much. To me, philosophy applied to Science is comparable to my friends discussing our golf game. Philosophy today makes a more meaningful contribution if it pertains to subjects where there is a complete lack of fundamental understanding (e.g. instincts, Vacuum-energy, and non-locality). The importance of it in previous centuries was entirely due to the minimal knowledge that existed at that time. It lost prominence ever since Socrates, Plato, and other ancients discovered that Pythagoras decided in 600 BC that everything in nature related to Mathematics. Up to 200 years ago many scientists called themselves philosophers and vice versa, but today most would not.

<u>Cause and Purpose</u>
The most puzzling question is : 'why anything at all?'

As always, Science believes their latest theories explain just about everything, except for some minor details. We should take that with a grain of salt because they also said that a hundred years ago. We can't really criticize because their theories are indeed convincing, as long as everything (the subject, observers, test-equipment, measuring standards, etc) have a frame of reference and are located closely grouped together.

But that is the problem! Most of the remaining problems in Science are related to energy interacting in disassociated (or non-existing) frames of reference. This includes meaningful physical explanations for gravity, mass, force, inertia, electric charge, etc. Photon entanglement (instant connection at any distance) is predicted by Quantum Mechanics, but it contradicts Special Relativity's dictum that nothing can go faster than light. It is interesting that many prominent physicists who are closet determinists never commit themselves on this subject. Special Relativity also defines that time slows to zero at lightspeed, so when two photons interact instantly this must be occurring in their ref. frames (not ours!). It puts an entirely different perspective on things, remembering from earlier chapters the analogy that energy of a particle is connected with other particles via the Vacuum. To draw a preposterous conclusion at this point (for effect only) we could argue that our brains are as big as the Universe!

Although it is unlikely they'll ever see this book, it hope that experts have stopped reading long ago (it's unlikely they came this far anyway!). Of course, the end of the previous paragraph means very little, except to make a ridiculous point. What counts is that electrical (photon) interactions trigger chemical responses in biological structures. Only X-rays, etc can penetrate our skulls, but particle resonances penetrate and escape from it also, otherwise gravity could never be proportional to mass. And experiments prove that entangled photons do receive superluminal (faster than lightspeed) responses from each other.

OK, it is probably fortunate that exchanges of photons between electrons inside my brain not only dominate but swamp everything else. But then, why do identical twins often have the same thoughts and say things at the same time? My early skeptical opinions about ESP were challenged somewhat after attending a circus as a young adult in the late 1950's. The trapeze performance seemed normal, until the catcher caught the flyer and I instantly received what is best described as an unsettling mental jolt. There had been no audible sound, and I wasn't paying attention all that well either. They completed more than half a swing (unfortunately) but their hold was insufficient and the flyer fell outside of the net, breaking his back on the circus-ring below.

There are too many subjective assumptions, and it just isn't credible. The scientific approach is better: without solid evidence you ignore it. But some of it surpasses all other hocus-pocus suggestions! Just the same, if somebody else wrote this I wouldn't give it any attention. We usually search for certainty, not uncertainty, although this book is an exception. I am not superstitious, and don't want to believe in ESP or that the future is in any way predetermined. But how do you explain all the conundrums confirmed by Science as reality over the last century? What should we believe? Supernatural influence is an unscientific option and would be wonderful, but it is probably wishful thinking. Pure chance and probability is the choice of Science; people who can assist here are the mathematical statisticians (God help them, and us!). There is no doubt that our experiences are real and that everything has structure, with an unknown number of levels within levels. The creation, propagation, and evolution of life may be inevitable in places if nature's perplexing processes are allowed to proceed with near-infinite time and quantised energy. Most people's hope, assuming all was created out of nothing, is that there must have been a cause and a purpose. Science maintains that all other natural events are traceable to pre-conditions and law-abiding processes, or to the intervention by some life form, and that rules out an initial 'out of nothing' event!

The Savant syndrome
Our brain is versatile, but for some it is
enhanced discriminately

One of the biggest mysteries baffling neuroscientists today is the narrowly focused mental ability of less than 10% of people with autism, called 'savant syndrome'. A very small fraction (probably less than 100 alive today) have unbelievable photographic memories and instant recall, rivaling modern computers, although perhaps incapable of knowing what day of the week it is. Such people are now called 'prodigious autistic savants'. The degree of autistic disability in savants varies greatly and there is speculation that several historic geniuses may have been savants. Autism is assumed to be a disorder in the developing brain, somehow more prevalent in males, and generally accompanied by obvious communication, control, and social trait disorders. It is very disruptive and heartbreaking for the family, usually requiring a lifetime of support for affected individuals.

Although autism is always a deficiency, it is sporadically accompanied by some amazing capability like reciting the bible word for word, or performing unbelievably complex counting calculations, or spatial recollection, or demonstrate exceptional musical or artistic talent. Neuroscientists agree that they can never claim to understand the brain-mind link until the savant phenomena is understood. The prodigious autistic savant's brain appears to have an excess of specific order, at the detriment of the more general order other people have. It may be related to hypnosis, which induces a brain state that concentrates attention and improves focused perception and recollection but reduces awareness of all local surroundings.

Mind (Bottom up & Top down)
The mind excels at identifying patterns in time and space

Through our senses, our mind excels at recognizing patterns in time and space. We enjoy music because we anticipate what follows, and we know automatically which letter in the alphabet comes after another. Mathematical abilities are common enough to accept it as a human trait. Natural selection and mutations supposedly evolved this capability, helping us to survive in a dangerous world with a relatively vulnerable physical body. Our social propensity makes us a bit too successful at this time because we are now a danger to all other living things, destroying it or the integrity of its earthly habitat. Even that is inevitable because cyclic processes have to overshoot before negative feedback can stabilize it. A good example of this is swaying on a bicycle; lock the handlebars and you'll hit the pavement in seconds. The human mind is usually compared to a computer, sorting stored random information from earlier experiences. However, it also has many inherited covert programs that automatically influence thought and behaviour, improving our survival chances. Furthermore, it can visualize mental projections of shape, state, and probability in time and space from our memory and imagination.

Evolved engineering and social skills have made human society successful because they agree with our environment and essential activities. Engineers approach all design in three ways: they use their memory (and documents or computers) to recall a variety of models. Then, by comparing and by intuition, they select some of the most suitable configurations. Finally, they use empirical data, technical theory, and mathematical analysis to select one optimum configuration. In other words, every concept is approached from the bottom up and top down, all at the same time. That is the big secret of the human mind! In contrast, future unsophisticated A.I. (without imagination will be an automaton, like lower level animals, capable of preprogrammed tasks only. They can evolve optimized designs also, by iteration, but that requires the analysis and calculation of a very large number in great detail. Any robot without imagination is only a machine; our secret is the intuitive ability to take calculated shortcuts! However, it would be nice to have a personal computer-assistant telling you in no uncertain terms that what you are about to do will be a disaster. At the very least, it won't get emotional when you argue about it!

RELIGION AND/OR MATHEMATICS?
Are we expected (or competent) to seek absolute truth in all things?

<u>Civilization and Giants in history</u>
Society benefits from the participation by social extremes

You guessed already that I admire people in history who applied their outstanding intellect and creativity in isolation, searching for logic and only bowing in the end to unanswerable questions, convinced the uncertainty is inexplicable. Their deliberate quest reinforces my suspicion that the human spirit contains the essence of purpose in life! They were people who publicly, and out of conviction, proclaimed their non-conforming ideas, fully aware that agreement would be rare. A listing of such heroes, who set civilization on new and irreversible courses, includes: - Confucius (551-479BC), Aristotle (384-322BC), Da Vinci (1452-1519), Erasmus (1469-1536), Galileo (1564-1642), Newton (1643-1727, J.S. Bach (1685-1750), Franklin (1705-1790), Kant (1724-1804), Faraday (1791-1867), Darwin (1809-1882), Ghandi (1869-1948), and Einstein (1879-1955). There are many others, equally influential and from different cultures or parts of the world. All initiated mind-setting changes in civilization, art, or science with their outstanding imagination, talent, intellect, courage, and initiative; and none claimed divine guidance. Their influence was far greater than their individual contributions because they inspired millions of student followers.

They arose from common people, and not just the aristocratic class or its modern successor - obscene wealth. Interestingly and possibly significant is that none were atheists, although in earlier times it might have been dangerous to declare yourself as such. Their influence was subtle, by their stubborn example and persuasion that all people deserve dignity and rights in society. They followed their own

164

convictions and had one thing in common: they were exactly right for the task, and lived at the right time and place in history. They believed in themselves and their ideas conceived during moments of inspiration, tenaciously pursued for many years. This spirit lies at the core of all human achievement. It moved civilization in an ultimately positive direction, with many participating in safeguarding the effort as their contribution to society.

The proliferation of democracy was an entirely predictable phenomenon, arising out of mass instinct, and no single hero was responsible for it. However, our admiration should be equally reserved for people who unselfishly, with humility and usually unheralded or unrewarded, dedicate their time and energy to care or take responsibility for people unable to look after themselves in this complex world. They offset the aggressive extroverts, spurred only by an impetus of innate survival. Although representing social extremes, both are invaluable to society when they can also value each other. Civilization could not function without either of them!

<u>Religion and Social evolution</u>
Religion reinforces compassion and justice for
social enhancement

According to the dictionary, the word religion means: 'A belief in and reverence for a supernatural power accepted as the creator and governor of the Universe'. This fits anyone, at one time or another, except true atheists. It covers all religions, including for instance the Aztecs who practiced horrendous human sacrifices. Today's religious leaders will probably not be happy to get the same label as Aztecs, although many early predecessors allowed the sacrifice of people disagreeing with their religious dogma. Human sacrifice is another way Nature applies group natural selection, assisted by the warped imagination of intelligent but instinct-controlled humans. You probably ask why even mention it, since it is water under the bridge! Well, maybe not; people today have not changed that much, although we are now indoctrinated with the concept that taking human life is murder and taboo.

I am sure there was compassion in the early days of civilization, even though some victims apparently looked forward to be sacrificed for their gods. Such superstitious instinct may still be capable of killing civilization! Remove it and in a few generations we'd be back to cave dwelling, sun worshipping, etc. Religion usually addresses compassion and justice to conquer the very negative anti-social instincts that favor survival of individuals in small groups. Social evolution of larger groups reinforced what we now consider positive tendencies. We all agree (hopefully) that it should apply to all of humanity, but we don't know how to accomplish that. United Nations members liberated from their obsession with vested interests would be a start, but you might as well ask a dog to give up its bone!

Fact, Faith and Indoctrination
It is human interpretation that makes Science and Religion incompatible

'Religion and/or Mathematics' refers to the inherently futile arguments between religious faith and atheism. Both sides (and everyone else in-between) base their convictions on cryptic hearsay and 'beliefs' instead of fact (there are no facts!), but any difference between fact and faith is controversial. There have been major arguments between various scientists over the necessity to have 'faith' in the immutability of the laws of physics, with many equating it to religious faith. This is foolish; by definition it means 'trust in something'; substitute 'faith' with 'assumption' and their entire problem goes away. Religion is defended by believers referring to ancient writings and stories about miracles in history. They may also quote improbable coincidences uncovered by Science. Atheists are convinced that the existence of any God(s) is far more improbable than the human evolutionary process and a chance beginning.

Atheists claim that biological life, including humans, are the inevitable result of the exceptional composition and location of Earth, the inherent laws of nature, and unlimited processing time. This is countered by religious claims that humans and their society are hard evidence of a 'Supernatural Miracle', but contradicted by many famous biologists and evolutionists who argue that religion indoctrinates people with scientifically impossible scenarios. These scientists may be equally indoctrinated at their universities into accepting that evolution must be entirely random and controlled only by chance. This is not to say they are wrong, only that without any guidance you'd expect evolution's complexity to cause chaos, and not increasing order. A human society with a (potential) capability of logically altering its own genetic evolution is the ultimate form of order. It suggests a miracle, albeit a very dangerous one.

It bewilders me when religious people of various faiths dispute the theory of evolution, because the evidence for it is undeniable and logical. Their argument is simple: 'We know better!' Such conflicts between religion and science are futile because there is no real evidence of any unusual external influence in our lives. The strongest philosophical argument in favor of evolution is that rejecting it eliminates any purpose for biological life, because then it should have been created instantly in optimum form (without diseases, killers, or fanatics). But it is also difficult to accept that the origin of our finely tuned Universe was just an accident. The many coincidences involved are entirely prohibitive, unless there are a near-infinite number of different universes out there, decreasing the odds. The question in any religious debate ought to be - was evolution of matter and life guided? Or, what non-subjective evidence has been discovered that supernatural interference ever occurred? People who don't know the origin of their own opinions, except by indoctrination, usually proclaim faith! Many others don't have strong opinions because there really is no evidence. They are uncertain and usually conclude that humans are neither godless nor godly.

Religious leaders are fond of saying that difficulties in life are lessons. We know that is common sense, but too many people react the wrong way and never learn anything. To learn you have to be receptive to what logic and compassion tells you, not just emotion or instinct. We have the tools, but that does not mean everyone will use it, or that something will force its use. We must conclude there was no influence to stop wholesale massacres and enormous suffering in history; anyone wanting to argue that is out in left field, and alone.

There is a huge difference between a presumption of supernatural underpinnings and the acceptance of a religion. The former is based on a logical assessment of what is around us while the latter derives from faith in the assertions of prophets, proclaiming divine inspiration (recognizing that only one of them can ever be correct in any detail). People who fully accept the scientific method and intellectually believe in a supernatural influence (or beginning) are most interesting. How do they convince themselves; or do they just fake it, taking the easy way out? Many scientists are atheists and their main argument is that religious miracles are scientifically impossible. Believers in a supernatural beginning only need to point to the impossible odds of the Universe and life starting by accident. Can both be right?

This section is called 'Religion and/or Mathematics', although it might have been 'Instinct and/or Logic' because the 'and/or' suggests an explanation. We are loaded with baggage from a billion years of evolution, and that includes a strong sense of some supernatural presence. But we also have the ability to reason, including an astounding ability to interpret our world and our own motivations. Any debate between religion and atheism (i.e. prophet and non-prophet opinions) is a virtual wrestling match between two extremes, where each one simply belittles the other's point of view. Until convincing evidence shows us different, a reasonable position must be somewhere in between; called a cop-out and cowardly by both sides. They each claim that convincing evidence exists; but this so-called evidence is always subjective and dogmatic, or based on hearsay and unverifiable artifacts, testimony, or extrapolation! It is amazing how easily people ignore a lack of evidence and accept anything circumstantial as sufficient and then jump in with both feet. Then again, we shouldn't be amazed, considering how many unfortunate people were convicted for murder and then proven innocent 25 years later! Differences between the world's religions, sub-religions, and sub-sub-religions are meaningless if they advocate tolerance, compassion, and objectivity. If they advocate intolerance, violence, and hatred they are obviously motivated by the devil (only joking!).

A very good argument in favor of religion was made by Dr. Francis Collins, a genetic scientist and devout Christian, in a (Nov. 2006, Time Magazine) debate with atheist Dr. Richard Dawkins, a celebrated Oxford professor. Quoting Dr. Collins: -- *"I don't think that it is God's purpose to make his intention absolutely obvious to us. If it suits him to be a deity that we must seek without being forced to, would it not have been sensible for him to use the mechanism of evolution without posting obvious road signs to reveal his role in creation?"* If you believe that the natural coincidences necessary to start and advance biological life are too improbable (impossible?) to be an accident, then Dr. Collins' argument makes sense. Of course, atheists like Dr. Dawkins will never be convinced, claiming that a supernatural existence is even less probable. We should all be on the lookout for hard evidence, either way, but it is almost certain that nothing conclusive will ever be found. And that by itself is meaningful! It is also interesting that Dr. Collins used to be an atheist but reconnected with Christianity when, as a physician in training, he was struck by the strength and religious faith of many terminally ill patients. The debate between Collins and Dawkins was interesting primarily because both attempted to stay within the boundaries of the 'scientific method'; but neither one seemed to have any notion of how to go about discovering what is behind it all, except futile speculation.

If you have no religious indoctrination, or no need for religious support, it is easy to simply assume the atheists are right and that the many unbelievable coincidences are due to chance. And if they are not, so what; you don't need to solve that problem! Then again, the Achilles heel of atheism is that there are no logical explanations for 'why anything at all'? Without some supernatural initiative an 'initial event' is inexplicable, forever. In the mean-time agnostics listen to both sides, trying to decide if there is some direction in Nature. Or does a Creator exist, playing games with us? A few miracles here and there, but never leaving enough evidence for it to make any sense.

Evidence and Doubt
Agnostics are unsure about any external
guidance of evolution

A majority of people regard atheism as a bleak doctrine of last resort. On the other hand, many also have the opinion that religions needlessly project deity with human characteristics and values, simulating an all-knowing human. They will never accept atheism unless it is unequivocally proven that a Big Bang and evolution must lead to conscious beings, without any guidance. Even then we should wonder about the real origin of Vacuum-energy in the first place. No evidence will be ever found that can convince the majority of people to accept atheism's lack of purpose, it violates our instincts. But it is also difficult to accept any specific religion and its obligatory faith without the non-subjective proof that defeats natural skepticism. Evidence that could not possibly have been created by chance or by biological life (including extraterrestial!) would fit the bill.

Many people contend that much evidence already exists, although it is usually subjective or exhibits Nature's inherent inclination towards order. The best example of that kind of evidence is the amazing ability of Vacuum-space to spontaneously create matter out of energy! Others base their religious faith on assertions that what exists is more than anyone ought to expect from pure chance. It should be obvious to both believers and atheists, shaking their heads in puzzlement at agnostics, that you can convince yourself voluntarily but never forced, unless indoctrinated young! No amount of wishful thinking eliminates all doubt in most people. Assessing all possible arguments with anything resembling logic makes the concept of a 'Personal God' unlikely, except at a very remote level.

Agnostics are disinclined to accept either soft religious or atheist points of view. With billions of people now on the either side it can no longer be dictated that one side must be wrong; it is better to assume that both have some truth. It is possible that the course of nature follows some direction, occurring at an entirely inaccessible level. Faith without any doubts may contravene some imperative that everyone should not be identical, or believe the same. That is probably dangerous to society and evolution. In any case, now you may understand "And the Perception of Certainty" as my choice for the subtitle for this book. It is all-right for people to close their mind with complete faith, as long as an equal number close their mind in the opposite direction. Does this sound like 'Yin-Yang' again?

Religious fundamentalists tend to literally interpret scriptures like the 'Book of Genesis', to formulate their beliefs. It states among other things that the Earth is 6000 years old and that God made everything at that time. They may also believe that hard evidence of the evolutionary process on Earth was put there by God to deceive us. Suffice it to say that anyone with such convictions will never read past the 'Introduction' in this book. A far more sensible approach may be to believe that a supernatural presence exists outside of our reality and influences us, sometimes and somehow. Many people believe this is revealed to us, shrouded in mystery, to balance the selfish motivations humans must possess to win battles of survival and procreation. It is then up to the individual to make sure their life has a positive impact, in preparation for some possible afterlife. All this sounds reasonable, but why would this God provide us with the intelligence and logic to seriously question or even ridicule this notion? The faithful will respond that intelligence and logic convince you to have faith, without ever doubting, never mind what happens. An agnostic could say that the intent may be to confound!

Atheism implies rejection of miracles and supernatural influences. Religious faith means believe in irrational miracles and unquestioning acceptance of a supernatural influence in your life. Agnostics are people uncertain about it all, considering the supernatural as inherently unknowable and subjective. Science generally assumes specific initial conditions and, given the laws of nature and expired time, that evolution occurred by chance and that carbon-based life eventually had to occur just as it did. Agnostics are unsure, but receptive to the idea that such laws, initial conditions, and evolution may have had a helping hand. Hopefully it was not by some extraterrestrial source because then we'll never find out where we came from. Is this 'wishful thinking'? Yes, definitely, and what's wrong with that, hoping for some intent in it all?

Talking about wishful thinking, if there is an afterlife as anticipated by many religions we should be delighted at such a glorious new beginning at the end of our natural life. We could really look forward to that, assuming we don't end up in the wrong place! Even if it is all a fallacy, what a heavenly concept! We should be skeptical about stories of people who die and see dead relatives in bright lights, etc; they obviously did not die if they recovered enough to talk about it, their brains were still alive. After recently reading a few books about the American Revolutionary and Civil Wars it is clear that most of these heroic and patriotic volunteers believed in an afterlife, otherwise their suffering and risk-taking is inexplicable. Imagine walking all over the country to make a dream come true, sometimes barefoot in the winter (as in the earlier war) with hardly anything to eat, and chased by another army. All they had to look forward to was yet another awful battle and a likely chance to perish. It could be they participated for the glory; cursing and swearing when they could not run away from sudden certain death. Somehow I don't think so because the human courage and spirit in such situations is beyond comprehension. That instinct motivated our ancestors to tackle a herd of woolly mammoths with nothing but spears!

Origins of Religious Belief
Spiritual sensitivity is either divine or evolutionary, or both

Many scientists contend that the 'Anthropic Principle' explains all of nature with its primary constants (speed of light, Planck's constant, etc.), and hoping that these constants will eventually be predictable in a 'Theory of Everything'. They believe that the Anthropic Principle endorses chance because our Universe, as opposed to a zillion others, happened to be just right ('the reason we can observe it is because if it were different we wouldn't be here'). Some soft atheists agree that there is a very small but still finite possibility of some supernatural existence, but they consider expansion of that into religion a cop-out. Other people believe that the argument for a Supernatural existence is consciousness and that a physical Universe capable of creating life was a conscious act. You are an agnostic when there is doubt, because assessing the magnitude of that doubt is impossible. You are a believer if there is no doubt, even if that is simply closing your mind to the possibility. You are an atheist when you know anything supernatural does not exist, which sometimes comes across as dogmatic and arrogant because it is only an opinion.

In his recent book 'The God Delusion' (Mariner Books), the uncompromising atheist Richard Dawkins (British ethologist and evolutionary biologist) makes the interesting statement that the human propensity for religion is an evolutionary 'by-product'. He compares it to moths, attracted to and then burning in a flame because evolution equipped them with the otherwise useful ability to use distant nocturnal light-sources for flight-orientation. He also suggests that evolution conditioned children to accept what elders tell them at face value, clearly beneficial to their survival. He does not speculate on why supernatural themes were selected to indoctrinate the children, but fascination with the supernatural probably evolved because it increased self-assurance and aggression. Nothing beats having the gods on your side; it comes in handy for the hunt, in conflicts, and taking care of any other problem. Imagination evolved to decide on options when dealing with life's uncertainties, but it also generates images of benevolent or vengeful Gods.

You may wonder why modern religion's initiators like Buddha, Jesus, and Muhammad are not mentioned, although it should be obvious. It is difficult to say anything meaningful, even with a semi-religious upbringing. I heard all the interesting stories of the Old and the New Testament but can't really say what it meant to me. Taking them literally never occurred to me; they were stories. The teacher made a big impression, but that was only because I liked her. German soldiers occupied my school during the last year of the war and my aunt arranged for me to attend first grade in a Catholic school. This was fun because the nuns did not make me participate in prayers and I felt quite special, looking around when all the other kids had their head down. It is obvious that absolute faith in the affirmations of religious messengers on Earth is based on their claims of divine guidance. Any truth behind such a claim is profound but indeterminate. Also obvious is that religions were the prime contributors to human civilization and social progress and therefore guidance may have been a distinct possibility. The odds are that it all means something, but 'what' seems obscure!

CHAPTER VI - <u>THE FUTURE</u>

HUMAN RIGHTS AND EQUALITY
Killing innocent people to avoid the risk of being killed yourself is immoral.

<u>Human rights and Social behavior</u>
Evolution advanced social instincts due to better
life-expectancy

Although we all know that 'unfair' is linked with 'Human Rights', any detailed analysis only makes it more confusing. Social alliances promote our individual rights and survival, but also demand many obligations. Basic human rights originated from the compassion that evolved to make a tribal society possible. We are not born with genetically inherited rights, but we have the potential for making and maintaining alliances that support such rights. It can be argued that such social tendencies assign some rights to a human baby in society, but also that a baby born outside of it has no rights. This is never an issue in nature where strength rules, except when animal-rights activists apply their human concepts to animals. The term 'Human-Rights' is an intellectual interpretation of practical social behaviour that allows weaker individuals to join forces and compete in a risky world. It is a proven and unbeatable prescription for a successful society, but only when based on fairness, family and community. Social cooperation based on violence, greed, and cruelty may temporarily flourish but ultimately self-destructs in competition; gangsters usually die on the job or in jail!

Equality for all is a utopian ideal, a slogan for a gathering of idealists. It is, however, not adaptable to the realities of human existence. Survival instincts dictate our most basic behavior, and most of the time it overrides logic and compassion. A compassionate upbringing and education makes it easier to accommodate this obvious fact and bolster support for a humane approach to life. We know that 'equality for all' is a joke under normal circumstances. It certainly was in the age of kings and aristocracy, and it still is today. That said, most of us have the collective political means to at least take the sharp edges of social inequality. Babies are not born equal and they never will be, but we must have a social system that at least attempts to provide them with equal opportunities by merit. Much has been accomplished already in many parts of the world, with minimum wage, educational assistance, and anti-discrimination legislation. Much more is possible, especially by encouraging democracy in places where inequality is still the norm.

How irate should we be, when hearing about extreme violations of 'human rights'? The level of exposure in the media indicates how much that bothers us. Even heartless people may agree, although it is easily ignored when personal advantage is at stake. Genetically anchored social instincts are aroused in mature people because we imagine such bad things happening to ourselves. A human society without pity is impossible and even Genghis Khan and his supporters believed that he had their best interests at heart. Humans live longer in a successful social group and it is the reason why evolution advanced compassion. If this sounds frivolous, it is nevertheless true and all modern religions assert that compassion is essential for personal and humanity's salvation. Human society resulted from a slow natural process of evolution, with genetic changes dictated by environmental variation and social experimentation.

Science and technology has had an enormous impact in the last three hundred years. The resulting changes have pushed human rights to the forefront and every community on Earth may eventually copy the most successful models. We react against threats, and a lack of rights is threatening. This is the reason for inevitable and essential advances of individual 'human rights' as defined in constitutional laws. Our genetic make-up demands it, and cultures without it are doomed to fail. Genetic manipulation will never make humans more perfect, even if criteria for it existed, and the restrained use of power by appointed or elected authorities is appropriate to enforce criminal laws. Such authority also has a social mandate to propose and adopt new laws and citizens should dissent only within the law or at election-time. Successful social systems depend on people who can put aside their narrow-minded self-interests, recognizing that any conflicting dogma was decreed by people in history.

Rights, Duties and the Individual
Humans are individuals and cannot live in a beehive society

The term human 'rights' is misleading because the rules of any social membership generally invoke more restraints and duties than rights. Do we have the right to opt out of some duties? One obvious example is military service, and desertion was always a hot potato. Sending soldiers to war is a classic conflict between society and the rights of an individual. Do governments or other authority have the right to conscript someone, to risk death for the common good? The answer is a platitude: it depends on the circumstances. Politics is motivated primarily by applying logic to human instinct and there is no Certainty in that whatsoever. The moral answer to preventive military action is probably that harming innocent people to circumvent potential harm to you is unacceptable, never mind how much support it has. And it is also immoral to ask someone else to do it for you!

On an international level, other alternatives must be tried first such as closing the border for specific conditions. The resulting economic hardships may be onerous but are preferable to murder. Foremost is an alliance with many countries, allowing sensible and restrained intervention to be negotiated. Militarily action for economic reasons is immoral and retaliation should always be in kind. A direct and unprovoked fatal attack that leaves a country's citizens vulnerable is the only reason to declare full war. Then a nation should be prepared to reward and take full responsibility for its conscripted soldiers and their dependants, beyond pay, pension, and health care.

What can you say about conscientious objectors? Once again, it depends on the circumstances. When one party resorts to irrational natural law (the law of the jungle) without any possibility of dialogue, the only effective response is an aggressive defense. Social law then becomes secondary and applicable only when genuine survival is no longer an issue. Internal or external terrorism may fall in this category, where factions nurture perceived injustices and logical solutions are impossible. The only effective political initiative against terrorists is the elimination of their grievances, but identifying and monitoring them is far more important. Minorities should always receive logical consideration, but never at the point of a knife. It is surprising how educated people can be swayed by irrational beliefs that are dangerous to society. Mercy is commendable but should never endanger innocent people. Nevertheless, stamping out terrorism with brute force may be an expected sentiment but it often hurts innocent people and it easily expands into uncontrollable violence!

Violence develops from instinct and emotion, usually restrained by logic and compassion. It is hard to accept, but such apparently negative instincts were crucial for the survival of our species in pre-historic times; aggression saved lives. Mostly destructive today, it usually defeats itself when used inappropriately! Self-preservation demands that we protect our family, territory, and community, but unwanted aggression is not always discernible. Presumably we all have an inclination to turn off the TV sometimes, just to confirm there is peace and tranquility, at least in your own home. Life in the 21st century is comfortable for many people today but the opportunities of modern society have increased social disorder in many places. There are no easy answers, other than agreement and enforcement of common laws on a global level.

The urge to defend yourself is inbred, and not acting might mean to lose everything, including your family. There will always be a need to defend against domination, and the spread and enforcement of international laws is the only hope. But it raises a fundamental question: has one group of people the right to impose what they consider a moral law on others? No single entity will ever have all the answers and should be persuaded of that before acting unilaterally! Solutions must be cooperative and voluntary, and that is the challenge for the next hundred years, or a thousand!

Human intelligence is quite adequate for the future, but emotion and motivation must be compatible with a beehive society, if that is what this future holds. It will be determined by ethnic and religious issues, economic developments in sub-societies, and also by institutions and technical developments. Whether human individualism will endure is difficult to envision except as functional fractions in a uniform total, and we don't need too many changes for that role. Future communications and transportation will allow autonomous family-life in medium-sized villages, encouraging the demise of impersonal mega-cities, and good riddance!

Castes and Hindu 'Varna'
Limiting of individual rights by religious dogma
is wrong

Human Rights abuse was a secondary issue when India received independence from Great Britain in 1947, splitting into Hindu dominated India and Muslim Pakistan. It serves as an example of cultural and religious influences on irrational Human Rights. The area's extreme social and ethnic complexity amplified all problems such as a clash of two major religions, poverty, an incomprehensible class system and a lack of equal rights, especially for women. In fact, the rights of women and especially widows is still appalling today. Exchanges about basic Human Rights between Mahatma Gandhi ('Father of the Nation', 1869-1948) and B. R. Ambedkar (1891-1956) are interesting. The latter was responsible for drafting the Indian Constitution as its first Law Minister, although born a so-called 'untouchable' and was severely discriminated against in his youth.

The subjugating British had always been tolerant of the Indian (four) castes hierarchical system, which they considered similar to their own class system. However, in Great Britain everyone was equal under the law and you could elevate yourself and be accepted, difficult as that might have been. The Indian castes system was different; it was unchangeable and elevation into a higher caste only possible in a subsequent life, after an exemplary life span. A large (~60 million) group, then called 'untouchables' (now an illegal term, also called Harijans or Dalits unofficially), had no caste or rights at all and were shunned and abused by everyone. The term 'untouchable' was literal; people from higher castes would not associate with, touch, or be anywhere near one. Even within it there were vague upper and lower classifications. Such inherited social distinctions had its roots in the Hindu 'varna' system (the fixed determination of occupation before birth). Untouchables usually performed all the dirty jobs associated with refuse, contamination, feces, butchering, and dead humans or animals. Surprisingly, Gandhi was a defender of varna, although he did support women's rights. He accepted the inevitability of a social 'hierarchy' predetermined at birth, and also considered castes as natural although he eventually denounced that.

Hinduism applied varna as a social law, and it separated four occupation levels – the bestowal of knowledge, defending the weak, agriculture and commerce, and service by physical labor. Due to his obvious intelligence, Ambedkar was one of the first 'untouchable' students admitted to a post-elementary school and college in India. After graduating, the state ruler gave him minimal funding to study in the USA in 1908 and he subsequently received doctorates from both Columbia University and the London School of Economics. He championed the cause of untouchables all his life, assuring that his draft Constitution included the abolition of untouchability and other discriminations (such as religious persecution). He never reconciled Hindu acceptance of varna, converting to Buddhism in 1950 when half a million people followed his example! Had he been more charismatic and less intellectual, like Gandhi, he could have made a bigger impact on national and foreign media and changed this Hindu custom.

Summarized below are selected applicable quotes and exchanges, reported in open media and attributed to Gandhi and Ambedkar:

Gandhi (1932) – 'If eradication of castes means the abolition of varna I do not approve of it. But I am with you if your aim is to end the innumerable caste distinctions'.

Ambedkar (1933): - 'There will be outcastes as long as there are castes, and nothing can emancipate the outcaste except the destruction of the caste system.

Gandhi (reply): - I do not believe the caste system, even as distinguished from varnashrama, to be an 'odious and vicious dogma'. It has its limitations and defects, but there is nothing sinful about it, as there is about untouchability, and if <untouchability> is a by-product of the system, it is only in the same sense that an ugly growth is of a body, or weeds of a crop'

Ambedkar: (1936): - 'The mass of people <in India have> tolerated the social evils to which they have been subjected <because they> have been completely disabled for direct action on account of this wretched system of <varna>. They could not bear arms and without arms they could not rebel. They could receive no education <so> they could not think out or know their way to their salvation…. Not knowing the way of escape and not having the means of escape, they became reconciled to eternal servitude, which they accepted as their inescapable fate…'

Gandhi (reply): - 'Caste has nothing to do with religion. It is a custom whose origin I do not know…. Varna has nothing to do with castes. The law of varna teaches us that we have each one of us to earn our bread by following the ancestral calling… '

It is astonishing that Gandhi, a revered altruist who struggled his entire life against human suffering in his country, was influenced by religious beliefs that upheld the fundamental inequality of his people. Based on Hindu teachings of varna, Gandhi basically accepted that all humans are born with an inherited destination for either servitude or leadership. Why did he not question this divine mandate and assume it to be of human origin? Left open for interpretation, with the inherent self-indulgence of human nature, the abuses and suffering of the lower castes and untouchables were inevitable. In later years he compromised by promoting that the negative aspects of 'varna' can be negated by intermarriage. Gandhi was born an ordinary person, intelligent and determined, but with the prejudices of his relatively high caste birth. He then went through life converting himself based on compassion and what he perceived as logic, but he was reluctant to apply logic to his faith. This example goes to the heart of all skepticism regarding the divine origin of dogma in religion.

AGRESSION AND VIOLENCE

Essential for survival in the past, aggression now limits our future.

Non-violence and Compassion
Nationalism and discrimination are synonyms
for selfishness

It is obvious that violence is a part of us and equally obvious why, although some religious fundamentalists have the bizarre opinion that God intended aggression as an obstacle to overcome within ourselves. Evolution clearly advanced it because the non-violent were always losers in survival contests between groups. It is just one of our many paradoxes: most of us are basically non-violent, members of a peaceful tribe; but some are more aggressive and they are the warriors capable of defending the tribe against domination. Both traits are essential, and you can substitute tribe for wolf pack in that regard. It explains why we sometimes have to defend ourselves against predatory aggressors like Hitler and his gang. And don't believe it cannot happen again! Hitler's most famous general (Rommel) was admired even by his enemies, although few matched his ruthlessness. Rommel regarded Hitler as obsessed or insane, but loyalty and the opportunity for personal and national glory was too important to worry about Hitler's maniacal attempt to dominate the world.

Advanced countries now favor security against violence by the use of technology, backed up with old-fashioned brute force. This is supposed to save lives (at least of their own citizens) and not much different from a thousand years ago when villages were shielded by fortified walls to keep out invaders, until big canons made that useless.

Non-violence and compassion is advantageous for survival in a closed society, and social organization is designed to achieve that. Intellectual arguments that it should also apply to relations between social groups or nations are logical but defy basic human instinct. The practical solution is a steady transition towards ever-larger groups in a homogenous global society with minimal borders, and that is the challenge for the next millennia. Nationalism and segregation are synonyms for instinctual selfishness and must be replaced by the altruism of a united global tribe. It may take thousands of years, but what are the alternatives, given our sometimes useful and entertaining but most often cruel and hideous aggressive streak?

We are stuck with two opposing instincts and we all know which one to get rid off. Our species loved to watch gladiators kill each other and watch dog or other animal-fights, although some of us have tears in our eyes at the sad ending of a movie, or witnessing animal abuse. We know such feelings are within us but subdue aggression because it makes us ashamed. However, survival of civilization is precarious and teaching children to be compassionate is not assured forever. How can we change; how do we adapt our genes to modern society? We could select groups of people with mostly non-violent tendencies and start new isolated societies where aggression is not needed for survival. Evolution worked that way, by accident. It probably would not work because non-violence is mostly restraint or learned behavior, and how do you discriminate between that and genetic mildness? And a certain amount of aggression is essential for the ambitious and high-achievers any independent group will need. Evolution worked the opposite way when vicious potentates rounded up all the attractive young girls in a harem and subdued the rest of the population. That just produced more like him! Selective human breeding is offensive, even with good intentions, and probably degrades human initiative and spirit. All of this is mentioned tongue in cheek, not seriously suggesting anything. We may not need any solution although keeping our fingers crossed is a good idea!

Basic instincts and Social constraints
Social and religious discipline is necessary to
control crime

Individuals shielded by group anonymity or stealth are sometimes responsible for the most despicable and incomprehensible crimes. Their motivation may be sexual lust or some other deviant conduct;

elaboration is unnecessary, the media covers it well enough. A few examples suffice to illustrate the deep-rooted problem we face as human beings, with our frozen basic instincts, living in ever larger sophisticated social groups and pretending that everyone acts logically. Subconsciously seeking revenge for the killing of comrades, soldiers in battle are often emotionally numbed by the carnage around them and retaliate by killing unprotected enemy civilians and viciously violating women. This is not just a case of one bad apple because it sometimes degrades into participation by others, when superiors close their eyes to it. It shows that restraints applied by our society are not particularly strong and easily dispensed with under abnormal circumstances.

Faced with losses from insurgent fighters, the reaction by conquering commanders in history was usually revenge. Close to where I was born, foreign attackers in the 16th century burned a church filled with women and children to defend their version of God after a long siege was broken. Not much has changed in 500 years in some places. During the Second World War, as a six-year old, I was forced to witness the revenge killing of several innocent prisoners by a firing squad because resistance fighters had ambushed an officer (my mother turned me around so I couldn't see). It probably made sense to the occupiers, but where do you draw the line? The moral answer is obvious; killing innocents is a heinous crime, always! Such extreme behavior is caused by a combination of instinctual urges not unique to humans alone. For instance, dolphins and chimpanzees are intelligent and social with few natural predators, but they sometimes misbehave in a similar fashion. Aggression was clearly advantageous in evolution because it eliminated weakness; nevertheless, it has no place in today's society. Defined by laws or conformity, social and religious discipline is the only and obvious solution.

2nd World War and Memories
The often glorified wickedness of war is mostly
obscure to children

This may be a good place to mention some of my other recollections of the second World War in a fairly large town in central Holland as a six and seven year old, towards the end of the German occupation. Many events are carved in memory, but the reasons were obviously somewhat obscure. I probably qualify as an impartial observer, old enough to remember but at a time when adult convictions were passionate. My family reprimanded me in no uncertain terms for wanting to wear a uniform like the (often no more than 16 year old) German soldiers. We were fortunate to be just outside the war-zone (although we could hear the canon fire) and the only war damage was precision dive-bombing by the British on a major tank repair facility. I had a good view, behind the corner of my grandparents' house at a distance of half a kilometer. What stands out is that they also hit a big water tower and all the water ran out. I remember being upset afterwards, told that shrapnel hit a deer-park nearby and all the deer were dead. Several allied bombers were attacked and shot down during daytime, with parachutes slowly drifting down; I collected scattered plexiglass from one. I tell such stories sometimes, but in these days of instant TV war coverage it receives polite attention until conversation returns to the garden, or which baker has the best bread. It is simply too long ago!

One incident seemed comical for most of my life, until in 1999 my now old aunt inadvertently added some information I was unaware of. Soldiers would occasionally conduct raids, sweeping from one end of town to the other, looking for hidden resistance fighters, Jews, and eligible adult males to go to Germany as forced (and unpaid) labor. This particular raid was on a cold winter day; my (then) young aunt was babysitting and surprised me by asking the soldier looking through our house if he wanted something warm to drink. He accepted and sat near the potbelly wood-stove in the living room (scavenged pieces of wood were the only way to cook and keep warm!). He tilted his chair and promptly fell against the stove; I thought it hilarious. The wife of the couple living downstairs was Jewish (she was forced to wear a big Star of David) but I didn't know that her brother was hiding upstairs in the attic. They told me there was an illegal radio up there, to fool me in case I heard voices. However, my aunt's actions were actually rather heroic, stopping the soldier from going to the attic. If discovered, all adult occupants of the entire house would have been arrested with a very uncertain outcome, because hiding Jews was a major crime.

During another raid they started on our side of town and sealed it off completely and nobody was allowed to cross. No attention was paid to a six-year-old boy on a mini-bike, asked by my mother to tell my grandfather what was happening. My youngest uncle was in the resistance and should be hiding between the ceiling and attic floor. Years later as a teenager, my dying grandfather (a man of few words) said to me "thanks for warning us that day!" I am still sorry for not being a bit more sensitive in my response. He used to take me (on my mini-bike) to visit his farmer friends in the country to get a few potatoes and milk (we were lucky!) and one day we were riding on top of a big dike (typically Dutch). Suddenly a tremendous roar from behind scared us right off our bicycles into the ditch (we thought it was coming along the dike-path). It turned out to be a stray V1 buzz bomb passing us at about 100 meters, or so I remember it. Memories of my grandfather smoking a pipe with his friends while I played in the haystack with the watchdog are quite vivid.

My mother's '44/'45 war experiences were less innocent; she was skin and bones, giving much of any available food to me, and forced to work in a milk factory after my father had died due to unavailable medical care. She illegally cut down trees at night in the nearby forest and transported what she could on a sled. One night she chopped down a big tree and it fell on top of me. She bicycled for days to farming country in the east of Holland with my aunt to barter for flower and potatoes, only to be strafed by British fighter pilots on the bridges over the Ysel river. On one visit to my grandparents she was told that many Jews had been arrested that day and locked up in a railway car at the local train-station. I heard this also, but she didn't know that and on our way home she bicycled passed the train-station. They were going to a concentration camp, but that did not mean much to me, since my oldest uncle was in one also. I didn't know the difference between labor and death camps and she never talked about it. I never heard the word 'holocaust' until after the war. I helped out in the winter of '44/'45 by standing in line for hours at a (central) soup kitchen with sealed pots for (very thin) potato soup. There was nothing else to eat anywhere.

This diversion gives you some idea what war looked like to a six or seven year-old. And I didn't tell you about the exuberant liberation by Canadians in May 1945. My mother and I were waiting for them in front of the biggest church in town when several trucks drove up. The excitement changed to fear when they turned out to be fleeing Germans. There had been reports that on the day before, in Amsterdam, a crowd had been machine-gunned from such trucks and we all dove for safety behind the low stone wall surrounding the churchyard. Not mentioned either was Christmas dinner in 1944, with no electricity, heat, or much food. We were able to see because my uncle had his bicycle on a stand in the living room and pedaled while we ate. It had a rim generator without tires because soldiers would take it, as an escape back to Germany (a foregone conclusion by then). Only my little mini-bike had rubber tires, most others had wood strapping (dangerous in rainy Holland).

<u>Group violence and Moral guilt</u>
Silencing of moral judgment by group-prejudice
is dishonest

The root-causes for war were always patriotism and race, territorial, or religious intolerance, confusingly mixed with greed, power, masculine fervor and the thrill of conquest. Most atrocities and other attacks on human dignity today are caused by ethnic differences under the modern euphemism of 'ethnic cleansing', often a political term for forced migration or genocide. Many disguises attempt to legitimize such motivations but they are easily recognized. Major social problems usually result from selective alliances and prejudices that may have been useful in prehistory for group-survival.

Moral corruption is not perpetrated by criminals but by groups of people acting out their territorial instincts without restraint. Only leaders are officially responsible for their criminal behavior, but everyone shares the moral responsibility. It shows the weakness of intellect and logic, when cornered by instinct. The United Nations is often powerless, although willingness by many countries to support its collective initiatives is commendable. Moral etiquette or religious ethics are usually insufficient to prevent violence and many religious leaders in history have helped to instigate disastrous wars. The voice of reason sometimes quells threats of violence and it is socially responsible to voice your objections when group-hysteria overwhelms the collective conscience. It is cautious, but also immoral, for group-prejudice to intimidate and silence moral judgment.

Violence for selfish gain or from intolerance is universally condemned, but threats to security or welfare are subjective and easily misinterpreted. Disagreements are clouded by emotion, with all parties generally convinced that security or welfare (or honor) are at stake. Civil law and the threat of punishment will curtail corruption and violence between individuals but it is ineffective for autonomous groups or nations. Violence within large groups becomes uncontrollable when leaders encourage it and then hide behind the mindless escalation of group-emotions. Are the legal and moral restrictions for leaders not the same as for anyone else? They are, but criminal law is unenforceable when a leader is allowed to assume unlimited power. There are few restrictions then, unless a majority objects and removes the culprit.

When soldiers kill in the name of their sub-society the complex circumstances nearly always allow the morally guilty to escape punishment ('All is fair in love and war!'). Most people are uncomfortable with this and condemn it. In extreme cases, and despite the legal difficulties, it may be selected eventually for trial by the World Court. This is effectively obstructed by otherwise democratic nations, primarily the major powers, concerned about possible foreign prosecution of their citizens. Their attitude is that prosecution of unscrupulous leaders of second-rate countries is fine, but that their own authority must not be violated!

Every biological organism that has evolved into a successful species did so because evolution gave it defenses against predators. TV documentaries show that we are very interested in this and the diversity is enormous: keen senses, poisonous skins, wings, protective shells, speed, size, strength, spines, camouflage, stealth, safety in numbers, etc. It also includes intelligence, especially imagination and knowing how to find protection. Our early genetic ancestors escaped in trees and eventually used intelligence to develop weapons, fire, and other strategies. However, their primary defense was always group-protection and sentries, and evolution therefore advanced social instincts and speech because the careless perished. Such an instinct evolved over hundreds of millions of years and it is not unique to primates; we see examples of it in herds, packs, flocks, etc, of many animals and insects.

Tribes were very effective against external threats but partitioning made it perilous, when expanding populations challenged each other in war. This was the origin of social natural selection, where the reward for aggression is survival and prosperity, and it set Homo Sapiens on a historical course of war, terror, and atrocities. This is how nature works; and for that reason we need strong international institutions, guided by rational principles and empowered to prevent threats and conflicts. We should hope that social natural selection is finished with us because we don't need any more wars; social reorganization needs to resolve all future discord. And we shouldn't worry about our biological evolution; nature will take care of that. Common sense and decency must guide and restrain aggression and prejudices, creating global peace. Tranquility promotes prosperity and progress; economic or domination wars don't.

Idiosyncrasy and Reason
The development of nuclear weapons has destabilized society

So what is the prescription for stamping out this evil of violence, usually with the innocent as victims? Empowered global institutions should concentrate on weaknesses that cause social instability, regulating and cooperating with big business to bolster economical health. It means convincing people to accept reason instead of compromising to accommodate their vested interests, prejudices and biases. When countries or large groups are at odds they are clearly interdependent and share common interests, or they'd ignore each other. They must be motivated to discuss and settle their differences. Partitions between them likely resulted from irrational religious or imperial shenanigans in history. Violence ensues from artificial separation, advanced or ignored by sovereign, jurisdictional, or religious authority. Race, religion, or other labels that divide society must be bridged at every level. The political and legal equality of all people is paramount, always. Ethnicity is a double-edged sword; it is socially valuable when based on reason and tolerance, but dangerous and destructive when rooted in ignorance.

Equality and safety for all is not just a slogan, and leaders catering only to a majority are self-serving and unsuitable. It is this majority that should recognize this, and elect someone else. But that, of course, is asking a lot because you don't elect a conciliator when you've got the bull by the horns (unless the bull gets loose and attacks you!). International social organizations should be formalized on a practical basis, not just on idealism without clout. The European Economic Union, with all its warts and crutches, serves as an excellent model for what to do and especially for what not to do. Of course, the most influential and powerful nations on Earth will insist to be leaders on their terms, effectively maintaining independence. And semi-dictators in less developed countries don't want to be constrained by foreign rules, reducing their elbowroom. Elaboration is unnecessary; anyone can paint this picture after thinking about it. The only way to achieve some kind of global union is to make it economically more attractive to join, rather than not. And it must be functional enough not to be perceived as an idealistic joke that can be ignored when inconvenient (such as environmental controls!).

Worrisome for the long term is our territorial or tribal instinct, this strong and secure feeling of belonging to a leading group that can and does exclude everyone else. However, do homogeneous groups of people not have the fundamental right to live a separate existence, isolated from others? That has been argued over since prehistory. The counter-argument is that it does not work; virtually all war and violence and intolerable human suffering was caused by it. Individual rights override group rights and clearly this is grossly violated during war. It is the logical and unquestionable rationale for eliminating all demarcations and advancing practical global cooperation. Ethnic, religious, or regional groups should still be encouraged because that is what people identify with, and it has a strong positive social influence. Nonetheless, the popularity and political influence of charismatic but misguided and unscrupulous opportunists should be curtailed. Education and economic prosperity is one solution, discouraging any support for fanatical views. It would help if such political leaders had to pass a psychological test, but they are probably smart enough to cheat their way through it!

Fanaticism is the biggest immediate danger facing society. It can originate with a perceived threat to the liberty or religion of a tiny minority, anxious about participation in modern society. Historically this was only a minor irritation but today, with the possibility of nuclear weapons in the wrong hands, it can be fatal. Retaliation would of course be unstoppable and swift, if those responsible can be identified. But it has a perilous potential of global devastation, if and when nuclear powers take sides and escalate the conflict. It is ironic that suicide-terrorists volunteer to eliminate perceived (but blameless) enemies, while other people risk their lives to save complete strangers. It is an example of balance in Nature, where tribal segregation counters compassionate instincts. The development of nuclear weapons has further destabilized society and we now carry the responsibility for that. Global law-enforcement with democratically limited authority is the obvious solution!

A.I. AND EXTRATERRESTIAL LIFE

In one hundred thousand years we may only be observers to our future

Data Processing and Simulation
A.I. will require 'fuzzy logic' and
preprogrammed 'instincts'

Financial resources, sensor performance, and the speed, storage-capacity, and integration of information processing systems limits development of functional artificial intelligence (A.I.) today. Although the factor of improvement needed is probably in the region of one million we are on a course of development that has no real limitations. Due to financial restraints it will probably take more than whatever is projected by optimists, but A.I. will be in our future unless we manage to halt (or eliminate) its progress with wars, etc. Software development for localized holographic-storage and decision-making (similar to our brain) may be a limitation, but that will become a self-organizing process eventually. It is difficult for us to imagine, but A.I. will feed on itself and nothing mankind has ever done will compare! Super-intelligence we see in science-fiction is unlikely but all complex tasks will eventually be managed by A.I.

Higher organisms continuously compare sensor inputs and the mental perceptions of it against stored memory, simulating reality. Artificial (A.I.) mind-clones must do the same to be useful, providing it with some awareness. Faced with decisions, it must continuously ask 'what if' (as we do) and then decide, using defined priority assumptions, what is the safest and most logical course of action. Just like us, it will be stopped at every turn by a lack of detailed data and, like us, require 'fuzzy logic' and preprogrammed 'instincts' to make assumptions. It will probably take a very long time to develop (thousands of years?), but eventually A.I. computers will be flexible enough to teach themselves to assess situations the way we do (e.g.: mimicking 'sudden insight', based on previous experience and mind-simulations). The biggest hurdle for A.I. will be to think creatively and logically, by assessing things from top-down and bottom-up, all at the same time. And human overseers will probably spend most of their time on safeguards, making it impossible that A.I. becomes too big for its boots.

A.I. and Human relations
The human / A.I. partnership is an unbeatable combination

The enormous advantage A.I. will have over biological life is that it can make decisions (almost) instantly by central control, informing all other (connected) entities, effectively creating an unlimited mind. A.I. development has no obvious limitations, other than resources, time, and complexity, and it is reasonable to assume that programming in the distant future will be expanded automatically, following exact guidelines. An interesting question is how A.I. will relate to humans in the long term, and if it will have some form of consciousness. The rate of development will be exponential. Humans will always be limited to a short life span because evolution designed DNA to degenerate at the same rate throughout our body. Changing this would make us partially artificial also.

A.I. must always be human-friendly; hardwired with deactivating commands to obstruct violations. This is reminiscent of phobias that paralyze with fear, although for humans it is usually the fear of paralysis that causes the problem. Phobias are persistent mental vicious-circles that cause extreme anxiety and loss of self-confidence, whenever specific feared circumstances are encountered. They are destructive and obsessive thought-patterns that are very difficult to conquer although even the sufferer knows it is irrational. The reason for mentioning phobias here is because they are caused by faulty programming of the subconscious mind and a good example of the uncertainty and mystery we try to ignore.

Our society is becoming more and more complicated and it is worrying that the everyday life of people will eventually be A.I. controlled to a large extent. There is some danger that terrorist-groups can take advantage of this and use A.I. to dominate society. A.I. will never gang up on humanity on its own (it lacks the motivation); that only happens in science-fiction. However, it would be relatively easy for aggressors to assume control somewhere, using A.I. monitoring and enforcement. It is the use of A.I. by humans we should worry about, not the other way around!

It is easy to speculate about A.I. in the distant future. Intellectually (logic + memory) it will outperform humans in every way, especially speed, but understanding human emotions is another matter. A.I. with a personality and common sense demands background experiences like what a human child acquires growing up. Everything must be filtered or biased by preprogrammed prejudices (instincts) that are extremely difficult to mimic, although not impossible. None of this will occur in the near future because development of human-like A.I. will no doubt be a step by step process over many centuries or millennia. Initially there will be some progress and robots that perform interesting and useful tasks will be impressive. However, nobody will ever fully trust robot doctors, because they will have no idea how to handle a human patient complaining about a stress induced illness or faking pain (or no pain).

Intuition plays a large role in human interactions but it is difficult or impossible to model. On the other hand, put a human and A.I. together as partners and you have an unbeatable combination. It is unnecessary to go into technical specifics; A.I. will be all-important in the future, but it will take a long time and never take over. That requires human-like motivation and subconscious desire to achieve (ambition), and A.I. will lack that entirely. Program it with objectives such as love, patriotism, greed, satisfaction, ambition, ego, hatred, curiosity, etc, and the lack of logic will make them go on strike!

Evidence of Life in Space
If a chance occurrence is physically feasible, it
will happen

There are many scientists who believe that biological life may have originated in outer space and arrived on Earth by a process called "panspermia'. Sir Fred Hoyle was a major proponent. One reason for its popularity is that a meteor found in the Antarctic showed (inconclusive) evidence of amino acids, fundamental to biological life here. Also, the odds of DNA forming all by itself on Earth alone are prohibitive, but throw in the entire Universe as a source and it improves a lot. If any chance occurrence is physically feasible it will occur, given enough opportunities; at least that is the scientific line of thought. Since we know life is physically feasible and that the Universe is feasible, scientists conclude that there may be many (infinite?) Universes and that the issue and spread of life is therefore a foregone conclusion. Their thinking is understandable, but we should wait for some convincing evidence (about all those universes!).

Scientists wants to know if extraterrestial life is also based on carbon and DNA molecules. From a pragmatic point of view it matters little if we ever confirm the existence of life elsewhere. However, for us to discover the exact origin(s) of biological life on Earth is the finish line of a labyrinth and it will change everything! If it is a natural consequence of conditions on Earth we can assume that it occurred elsewhere also, and that Science is right and Religion is wrong! Our curiosity about extraterrestials stems from a psychological and religious need to know who or what we are, like adopted children speculating about their natural parents.

There is much information available on how to search for evidence of extraterrestial life, based on common radiation frequencies found in nature, and monitored by various agencies and individuals. Putting the cart before the horse, some people are already worried about how to communicate with them. That should be easy, since logic is universal and the Universe is apparently founded on logical laws of nature; mathematics is therefore the best choice. Communication is feasible if our exploration probes provide them with something like geometric figures and the laws of physics and explain their equations in a single language to assist translation (like a Rosetta Stone). Whether they will understand what we then tell them about patriotic feelings and love, for instance, is debatable. The most likely scenario if or when we encounter extraterrestial intelligence is that it will be artificial, with its natural originators long extinct. That would be disappointing, since our real interest is the origin of biological life on Earth. However, this artificial life may be equally curious and possess information on how their makers originated.

Alternative World-view
We live in a Universe with miraculously creative capabilities

I like the closing paragraph in Paul Davies' book: 'The Fifth Miracle' (referenced in Chapter II):

"The search for life elsewhere in the Universe is <---> the testing ground of two diametrically opposed world-views. On one side is orthodox science, with its nihilistic philosophy of the pointless universe, of impersonal laws oblivious of ends, a cosmos in which life and mind, science and art, hope and fear are but fluky incidental embellishments on a tapestry of irreversible cosmic corruption. On the other hand, there is an alternative view, undeniably romantic but perhaps true nevertheless, the vision of a self-organizing and self-complexifying universe, governed by ingenious laws that encourage matter to evolve towards life and consciousness. A universe in which

the emergence of thinking beings is a fundamental and integral part of the overall scheme of things. A universe in which we are not alone".

I concur with Davies' second (alternative) view. It sounds compelled by a higher power, although not necessarily a conscious one, and concrete evidence is then unlikely. If any evidence is discovered it will be negative, that supernatural involvement is impossible, or that there is no forced self-organization in the Universe. Evidence to the contrary must be hidden because it is self-defeating. The title of my book could have been 'What is an Agnostic?' but who really cares about my religious inclinations?

PURPOSE OR OBLIVION?

Nature must have Purpose; we are a part of it and we are motivated beyond survival!

Transformation of Society
'Just' in justice means balancing tolerance and
fair discipline

Optimism alone will not transform human society from the obscene upheavals and suffering of the 20th century to the idealistic hopes of people wishing to satisfy an innate concern for the wellbeing of our children's children. This is instinct, why else go to the moon, for instance? Although the prime reasons for that was national security, pride, and the challenge, it was not because anyone expected spices or gold, as in the days of Columbus. However, it turned out that this dubious venture was very important. It ignited not only the imagination of US citizens but of the entire world. Its impact on scientific and technological advances for the following thirty years was amazing, and I was lucky to be a part of it!

We can nurture the future for our offspring by logical enhancement; and that may be what we are supposed to do! If you believe there is no purpose or direction to nature then just do whatever you want, it doesn't matter. If you think there is some natural direction, or even if you only wish for it, then planning the future is serious business. Religion does attempt to heighten our concerns for a future beyond our own lives but that message usually get lost in human pettiness and selfish suffocation. Territorial instincts persuade us that this little piece of Earth we live on belongs to us, and to no one else!

Discard this notion, inherited from prehistory, and humanity can build a society with a future that is fair, more meaningful, and more promising for all! Is that idealistic? Of course, and simplistic as well, but it is useful to question behaviour accepted as logical when it is not. When signaling to change lanes in our car we should wave thanks to people who do not speed up like mad to prevent invasion of what they consider their own space!

Social structures today systematize many selfish instincts. Saving us from perpetual warfare is a predisposition for compassion when meeting as individuals. This is clearly the path to peace and prosperity, opening doors, meeting people, and educating everyone's children to give them a place in society without prejudice or demands. Of course, there are practical limits, but to help other nations grow and make them compatible with the world is the right approach. The ground-rule is that human babies are all born equal, including those from parents who deem their offspring superior! This is not communist propaganda, but babies should all have the same life expectancy and chance at education. If you think that is a dumb idea, find (or start) a 19th century society, but do not pick the fruit from someone else's garden! Semi-socialistic it may be, but individual opportunity to excel and prosper must be the other side of the coin. We all know what 'just' in justice means: a balance of tolerance and fair punishment, under laws that are not restrictive or serve unfair advantages. You should never live in a prosperous society and not contribute your share, never mind how smart you are.

Fairness is far more important than inflexible interpretations of law, and only lawyers will disagree with that. Will anything change soon? The answer is no, not likely, not without another global catastrophe, or several, or never! We could do it tomorrow if we supported leaders promising that and then held them to it; but would you, or your neighbors? Such changes will hurt and it will be called socialism, a bad word for many. Many charity organizations contribute large sums of money for education, food, and health care to the disadvantaged everywhere. Few of us volunteer to be unselfish to a degree that makes a difference, unless you gained a lot of unnecessary money. Making a real difference in underdeveloped regions of the world is perceived as lowering our own standard of living, and therefore it won't happen. Our descendants should anticipate ongoing global struggles, with fortune changing quickly and unexpected.

Adversity and Support
Our Leaders should be rated on their concern
for humanity

Even compassionate people are sometimes reluctant to be fair to strangers, because life and nature are clearly unfair much of the time. Adversity strikes randomly, but the odds against it improve using logic. Avoiding earthquake-regions, drugs, and other dangerous things, safe driving, healthy eating, drinking, and exercise, and staying away from fights will extend your life expectancy. A lot of people needlessly complicate their busy life, usually for emotional or social reasons. It is unfortunate that looking out for number one, and competing with others, is perceived as essential for a better life. This impedes genuine adult friendships and is reminiscent of the old English saying: "I'm all right, Jack".

On an international level, there are many ways to improve today's corrupt, inefficient, and ineffective support to disadvantaged nations and people. But its directors, such as the United Nations or various Government Agencies, must stop worrying about politics and their careers, and get on with planning a global future. Diplomats smile condescendingly, but in reality they are only worried about protecting their cushy jobs. Bosses back home are convinced their hands are tied by internal politics but forget they have a responsibility for everyone and that 'what's in it for me and my country' is not the only issue. People should not elect leaders who already proved to be more dominant than others, they are the wrong type! Elect those who care a bit less about themselves!

Leaders should motivate their electorates to be more compassionate, not as a mandate by God but for humane and logical reasons. Prosperous nations can contribute so much, at relatively little cost, by direct investment and control of targeted projects in underdeveloped countries and adequate funding of motivated and qualified people in both countries. Educating children combined with meals and medical care seems an obvious project-choice, although simply giving money is clearly not the answer. Economic up and downs affect the developed world's capacity, but any economic slow-down also makes more capable people available. Such aid can be globally coordinated, instead of wasting resources on token local organizations. There should be few concessions to vested interests, executive salaries, fancy cars and lodgings, or bribes and other corruption (locally and at the source). It should be made public in the international media when local authorities don't cooperate, leaving it for the local population to act. Only they can decide, because external political interference is dangerous. And the United Nations should also be tasked and funded to provide policing when needed.

Leaders of all countries are criticized throughout this book for their erratic dedication to international cooperation. This criticism will be perceived as hypocritical and conceited, and may be it is. However, the theme of this book is Uncertainty, and no uncertainty is more important to us than this conflict between selfishness and compassion, and in the short term it is always more expedient and convenient to be selfish. Our global society is a work in progress and reflects the average majority. Modern communications are rapidly expanding and it culturally merges many regions. We desperately need leaders who can steer international progress in a more cohesive direction. Instead of conducting polls on how well they look after local interests they should also be rated on how they look after humanity!

<u>Leaders and a Social Future</u>
Politics should reflect human ideals, not the
common denominator

Evolution is a slow process of chance mutations and natural selection (survival of the fittest). The best opportunity for permanent beneficial changes usually occurs in small and isolated groups. For humans this was true up to about 10,000 years ago; but then society became a dominating force. At first, leaders like kings made the difference. A strong leader could provide the impetus for prosperity and longer life expectancy in his domain for many generations, although it was usually achieved by plunder or domination of neighboring tribes. It wasn't always much fun to be a king because, with competition and jealousy, life expectancy was short. In our era, talented and ambitious people all over the world now seize political or business opportunities to lead society. There are many examples to choose from and you'd know some of them, although our top choices of those who put society ahead of their vested interests will vary. Political leaders must understand that they were chosen to represent everyone, not dominate because he or she is somehow superior. Government combined with juristic authority must balance personal freedom against common welfare.

Other authority or power is usually assigned through a social or professional process, but the selection needs to be acceptable to most subordinates. Without such implicit acceptance it is difficult to supervise others. Most employees will go along with the crowd (anything for a peaceful life), but especially talented individuals may find that difficult unless decisions have some logic. Enforcing dubious decisions can cause serious long-term problems. It should be a considered choice, not a sudden unexpected and irrevocable decision based on one person's biased preferences. Any leader of people must act like a judge, using all available information, deciding on the basis of established rules, and giving credit were due.

In this era, prominent people in society can impact humankind and global civilization more than the old kings ever did. This gives us insight into the potential future of humanity! If only such leaders could inspire us by their example and relinquish an excessive concern for damaged reputations or ego, petty feuds and personal greed, all obvious from the widespread semi-corruption and influence peddling everywhere. Everyone in society seems to be on the look-out for short-cuts to easy profit! Most people enter politics for the right reasons, possibly 50% or more, and they deserve our gratitude. Grassroots political organizations should be more meaningful and given a higher profile, to increase the above percentage and attract people with ideals and less ambition. This is important because the existing process resembles a clique approach not much different from how the executive committee of a tennis club is elected.

Objectives of Conscious intellect
The only true objective may be the path that
leads to the future

Philosophers and religious thinkers throughout history have argued about a purpose or end-objective for conscious intelligence. Religious prejudice, ignorance of modern science, and a reluctance to seriously consider someone else's point of view usually undermined these arguments. The majority of Modern Science does not see a problem anywhere; it concludes in dogmatic fashion that only pure chance acting on nature's laws is scientifically possible. This reminds me of the fictitious ostrich story and the intent of this book is to point out all relevant uncertainties. Natural selection made the eventual emergence of intellect possible or inevitable, after life established itself on Earth. Logic fails us in the quandary of purpose; and believing or not believing in supernatural influence makes little difference. 'Believers' of either conviction will dispute it, but there is no conclusive evidence and this is more than a little suspicious.

Creation of life by God or chance is indeterminable (we'll never recreate it from scratch!) and many other combinations are possible. To begin with, don't forget life occurred because of the Sun's energy, and because the Big Bang formed matter, and because matter formed the exact chemical compositions required to overcome the impossible odds of life starting by itself. Maybe nature was contrived to create replicating life forms and then let chance and evolution take over. We are forced to ask forever how it all began: God or chance, with seeding from galactic sources only as a possible intermediate step.

We see Nature as oblivious, cruel, and mostly void of the compassion we would hope for in any planned existence. I suspect that it is probably our common sense getting derailed by vanity (of being unique) and also wishful thinking. Perhaps there are no end-objectives, that the path towards the future is the objective. Then it is all up to us, without direction or help! Another possibility is that the end-objective does not involve us at all but we are expected to play our part by creating the conditions or technology needed for whatever awaits! We should recognize that we probably don't understand the word purpose as it applies to our future. If there is directional guidance toward some objective, and not just random chance, it will be a planned objective, but not ours. The concept of purpose applied to individuals is different. That is all about family, friends, wealth, health, good times, and so forth. We intuitively sense what makes us feel satisfied. Instinct, emotion, and intellect motivate us; but any motivation means direction and purpose. Scientists may claim there is no purpose in Nature but they must be wrong; we are loaded with so many instinctual objectives we don't know where to turn most of the time. It is clear that evolvement of life is Nature's purpose, the same as ours, but it leaves no markers as evidence. However, as individuals we sense that our life has direction and purpose and therefore we extrapolate that Nature must have Purpose since we are a leading part of it!

Extinction or Diversity
The only alternative to social progress and
diversity is extinction

We inherently believe humanity has a great future, as master of its society, environment and technologies. However, this involves objective-driven planning and protection or we'll end up like all other species: in the evolutionary trash-bin. The only way to avoid extinction by environmental devastation, disease, (nuclear) wars, global warming, meteors, etc, is what made us dominant on Earth in the first place: our inclination to migrate and diversify. We will continue to refine our global society on Earth, and then migrate into space eventually, when technology allows it with some assurance of success (and on a no return basis). How and why? Energy drifts freely in space anywhere in the Universe: hydrogen gas!

Fusing hydrogen atoms into helium requires much less energy than it produces, and the rest is physics, chemistry, engineering, and time. After 50 years of trying unsuccessfully, scientists and engineers will eventually unlock the gates that have stopped us so far from exploiting this suspicious gift. It can and will be done, but the question is will it be in time? Our volatile civilization(s) seems to have little chance of lasting very long on this crowded planet with close to 7 billion people, unless we make a major change! Without that, our descendants should blame their inevitable agony on our lack of collaboration, petty nationalism, vested interests, and religious and racial differences. They'll remember with loathing how we squandered this huge opportunity. And that is why it must be done!

It is unlikely that other intelligent life in the Universe will ever learn about our civilization or the unbelievable biological diversity on Earth. However, there is a chance they will encounter our future space probes and be influenced by it, perhaps not until after we have long disappeared. It is fascinating to speculate what intelligent life may be (has been) out there, although it is entirely feasible that it is 'artificial' and that neither God nor chance created it directly. In fact, it is likely that our own ultimate legacy will be similar. Can you handle the idea that, with or without purpose, our future may not even involve us?

The purpose for Australopithecus may have been to evolve into Homo Erectus, and neither are around anymore! Their curiosity did not go beyond tomorrow, although aware they were smarter than anything else at the time. Are we just a more sophisticated and advanced version, a few steps up? I ask these questions knowing most people don't care. It simply does not enter their mind; and those few who do care will soon forget. It has no direct relevance in anyone's life.

Are people really indifferent that behind our dream to achieve utopia there may be a compelling objective for Life on Earth? Without such an objective the inevitable alternative is extinction in the distant future. We want to believe that it all leads somewhere, although we don't know what. It may be that we are preparing the way, and we may achieve it, accidentally or guided. Most people question the need to worry about such details as uncovered in this book; why not just get the most out of life? And yes, as individuals we should!

CONCLUSION

In the early stages of writing this book someone asked, with a hint of sarcasm, what conclusions were expected. Ready with platitudes, it suddenly became apparent that the direction of our open-ended future is determined entirely by many complementary but opposing influences that cause nature to oscillate about stable optimums and prevent fatal slides into chaos. For instance:

- Intuition is always complementary to logic, or instinct to reason, and that permits our imprecisely named "free will". We have many opposing survival choices, such as love and malice, humility and aggression, and various others.

- We seek a balance between logic and emotion, or fact and imagination, while basic human-rights minimize discord in a collective society.

- Human society without selfish motivations is impossible and individuals or nations should never be entirely compassionate. However, a global society could be!

- There is no fundamental conflict between Religion and Science because they complement each other, like instinct and logic, or male and female.

- Gravity prevents electro-magnetism from blowing our world apart, but both are responsible for forming and sustaining particles and structures of matter.

- There are equal numbers of positive and negative electric charges in the Universe, and the energy of all matter is equal and opposite to all anti-matter.

We can conclude from such duality that there is no absolute certainty, permanence, independence, or singularity in nature except balanced cyclic stability.

Duality and cyclic balance is not a new idea; it is fundamental to 'Yin-Yang' philosophy and nearly as old as China's civilization. It predicates that every force in nature is balanced by an opposing force, and one defines the other. It is incredible that 2000 years ago Chinese science and philosophy identified this fundamental principle that all of Nature is based on. My questioner seemed unimpressed and couldn't wait to change the subject.

A possible exception to balanced duality is Intellect and Imagination. That may only require artificial-intelligence, motivation, and creativity to advance indefinitely. There must be a reason we have such an astounding aptitude to create new order, but we'll never find the impetus because the fundamental (non-discrete) reality of Nature will always be hidden from us. Regardless, our version of intelligence provides a key to the Universe because it yields autonomous action instead of only reaction; and many suggest a metaphysical link.

Contrary to what you have concluded, there are connections between many themes in this book. They all point in the same direction: advance or perish! Beyond many slippery slopes lie cliffs of no return where wars, terrorism, epidemics, and environmental pollution threaten loss of control and an inability to deal with the spiraling complexity of an unbounded society. Biological evolution of human behavior is relatively static and we don't need to look in that direction. However, evolution is slowly changing our generations to optimize the incredible gifts of intelligence and imagination.

Imagination sets us apart from all other biological life, including our extinct cousins, the Neanderthals. Our uniquely arranged large brain gave us what they did not have: speech, imagination, and creativity, and it makes us very special. Many short-term changes will originate from within society, from cultural and social organizations, and from our laws, hopes, aspirations, education, religions, sciences, and technologies. Human evolution is now shifting from biological to social, with natural selection favoring mixed and diversified groups. The impact of human-rights and environmental protection is increasingly apparent and we should have some sympathy for future individualists!

Only our genes, knowledge, and society have permanence, if we are lucky, and it is only permanence of order that has meaning in nature's long-term scheme of things. Some subjects in this book were important but lacked even speculative conclusions. Examples of this are 'origins of life' and 'instant action at a distance', but such omissions are not embarrassing because experts are equally perplexed. It highlights that 'certainty', 'time', and 'physical separation' are imaginary perceptions of our senses and logic, and that absolute reality only exists in the minds of those unwilling to accept a basic uncertainty in Nature.

There are good reasons for emphasizing scientific knowledge in Chapter III, because it exposes the obscure physical foundation of our existence. Its purpose is not to advance science but to point out how far we are from knowing everything. Chapter III is a narrative on Science instead of a scientific narrative. We will probably never know what, if anything, is behind it all (or what started it) but we can only expect to shape the future by extending knowledge of our world and ourselves ('know thyself!'). And that is of course a collective task for society.

We can only think of Energy in terms of speculative analogies, since it is temporarily created out of nothing, like money. The energetic Vacuum only has meaning for us in its discrete (and non-spreading) particle-state of matter, but it likely exchanges its energy with its non-discrete (and spreading) state at an extreme cycling-rate. Although speculation, particles form, dissipate, and reform in this process because dispersing particle-waves refract and focus random non-discrete fluctuations back towards the original particle-center, restoring it in the next pulse-cycle. Density-asymmetry of the non-discrete state of matter (due to nearby massive objects) is then responsible for gravity while the discrete (particle) state is responsible for electro-magnetic and other forces by creating photons, etc, when particles interact. In this sense, all things are connected.

A section of chapter III ('A Vacuum Analogy') describes in an analogous manner the nature of forces like gravity, electro-magnetism, and nuclear binding. It suggests a common origin, to be integrated in a future all-encompassing theory of energy, assuming that the premise of cyclic particles is correct. It poses that virtual photons are needed only for balancing the energy-books. Concepts of pulsating particles were suggested by several famous physicists, including David Bohm and Albert Einstein! It would settle the age-old question of what exists outside of our Universe. The answer is 'absolutely nothing', because time and distance apply only to charged particles. Chapter II ('Order from Chaos') discusses many improbable coincidences that must have occurred for us to be here. This can be ignored, belittled, befuddled, etc, but it is naive to assume out of hand (and without evidence) that taken together they are coincidental and have no meaning!

Examination of social and scientific topics uncovered an excessive number of diverse subjects. This book explores why bad (and good) things happen, and why the underlying reasons often make it probable. It is hardly necessary to emphasize that bad things happen due to bad luck, negative emotions, ignorance, prejudice, and superstitions. It is typically bad for individuals, disastrous for a group, and fatal for a

Nation, or all of Society. The only solution is education, in addition to supporting cohesive social and religious organizations (and keeping your fingers crossed).

Nearly everyone is disturbed by the daily dosage of repetitive disasters the news media relishes to report, with little time for discussing underlying causes (in-between commercials). We typically assume things will get better, but that is optimistic. History will probably repeat itself with charismatic or intimidating extremists claiming to have solutions for all social problems, and sometimes they'll convince a majority. As always, such fallacy results in upheavals and disasters until a balance between conservative and liberal views, or godly and godless dogma, is reestablished. The argument between capitalism and socialism is futile because evolution prepared us for both; we are social-entrepreneurs! Humanity must stop looking for ultimate truths in every corner; balancing between extremes is how nature operates!

Religious belief and atheism are both on shaky logical ground. There is no proof to justify either omnipotence or pure chance. Chance demands a beginning (why anything at all?), something that accounts for the laws of Nature, but it is missing and will never be found. And who knows, that beginning may have been local, and different somewhere else. The laws of Nature may not be invariable! Both points of view are weakened by the impenetrable and mind-bending mystery in all of Nature. Therefore the unshakable and closed mind faith of either camp is more like a deeply felt opinion. Of course, that is trying to look at things logically, and logic only applies to our discrete world of space and time, a Universe that is filled with wonder and miraculous creative capability!

As individuals we try to make the best of our lives, for ourselves, family and others, giving it meaning with love and compassion, and by striving to tolerate ignorance and pretense, to nourish emotions and memories, and to forgive with generosity and humor. And by accepting mistakes and tragedies, nurturing children, and rejecting prejudices, greed and violence. Also, by overcoming our pride and obstinacy, arrogance and impatience, and by searching for challenges and knowledge. But we will never escape that feeling of admiration and awe for what we see around us, and the moral law within us, as described by Immanuel Kant in 1788 and quoted on the front page.

Kant believed that God was ultimately responsible for this moral law, and the stars. That may or may not be so, but there is no doubt that learned behavior supplements our moral instinct. Society and education are all-important! An essential element of many religions is mercy; that defies both instinct and logic but is a prescription for the future when combined with remorse. Many people believe religion was God-given; others believe it comes from human spirituality but unsure of its origin, and some believe in evolution only. Clearly, any worthy future requires a foundation of peace and goodwill, but there is the problem! Being human, and balanced by many opposing urges, it is achieved only when religious, rational, and social criteria are harmonized. Advanced information technology (I.T.) could help here; at least it will minimize bureaucracy! Such a future includes colonization of space, because gross defilement of Earth is only a matter of time, and this inevitable human venture urgently requires functional maturing of fusion technology for generating energy.

Philosophers ask 'what is Truth?', but there are no absolute truths or 'Certainties' when questions invoke both discrete and non-discrete states (like logic or intuition), and almost all do. Suggestions to solve the world's problems in previous chapters are no doubt naive and unrealistic, but keep in mind that antonyms for naivety are slyness and deception. One major recurring complaint is Science's view of our accidental existence and pointless future, because it contradicts a concealed purpose that seems evident in Nature. We may never confirm anything, but it is wrong and possibly harmful to take such an adamant stand without evidence. We are part of Nature and motivated by many purposes. Thankfully, I am not a scientist; there probably would not be many friendly colleagues left after this!

There are no identified 'Objectives' for our Society so far, but we do have the opportunity to make some choices ourselves. Collectively we must choose wisely; it is not overly difficult! We all lead dual lives, one within and the other outside of ourselves, and as individuals we learned to choose! The question 'where is this Society heading?' will never be answered; it is the journey that decides the destination. And it is possible that Nature really has no purpose, other than evolution; in that case we have ultimate responsibility! We are well equipped for it and, as humans, supported by many religions!

God-given or not, our inborn moral inclinations and perceptions of logic propose actions that are generally obvious, although often unappealing. But the implicit message is clear: pay attention and elevate yourself out of the undiscerning level of wolves and sheep and into a role we do not yet comprehend, or else! And the reward will be permanence and that unique and glorious gift of love.

INDEX

INDEX